T0139870

Advanced Information and Knowledge Processing

Information systems and intelligent knowledge processing are playing an increasing role in business, science and technology. Recently, advanced information systems have evolved to facilitate the co-evolution of human and information networks within communities. These advanced information systems use various paradigms including artificial intelligence, knowledge management, and neural science as well as conventional information processing paradigms.

The aim of this series is to publish books on new designs and applications of advanced information and knowledge processing paradigms in areas including but not limited to aviation, business, security, education, engineering, health, management, and science.

Books in the series should have a strong focus on information processing - preferably combined with, or extended by, new results from adjacent sciences. Proposals for research monographs, reference books, coherently integrated multi-author edited books, and handbooks will be considered for the series and each proposal will be reviewed by the Series Editors, with additional reviews from the editorial board and independent reviewers where appropriate. Titles published within the Advanced Information and Knowledge Processing Series are included in Thomson Reuters' Book Citation Index and Scopus.

More information about this series at http://www.springer.com/series/4738

Themistoklis Diamantopoulos ·
Andreas L. Symeonidis

Mining Software Engineering Data for Software Reuse

 Springer

Themistoklis Diamantopoulos
Thessaloniki, Greece

Andreas L. Symeonidis
Thessaloniki, Greece

ISSN 1610-3947   ISSN 2197-8441   (electronic)
Advanced Information and Knowledge Processing
ISBN 978-3-030-30108-8   ISBN 978-3-030-30106-4   (eBook)
https://doi.org/10.1007/978-3-030-30106-4

This Springer imprint is published by the registered company Springer Nature Switzerland AG
The registered company address is: Gewerbestrasse 11, 6330 Cham, Switzerland

Preface

Our era is characterized by constant digitization of information and major technological advancements. Software is practically everywhere: web, mobile, desktop, embedded, or distributed; software is developed to assist people in achieving their goals easier, faster, and more efficiently. Even traditional sectors such as banking, health, and administration are transforming their business model into a software-based one, in order to cope with the new business regime. However, building a business over a software product or service is not trivial; the software landscape is highly volatile in technologies and markets, and often unpredictable, forcing software companies to act quickly, fail fast, and adapt faster to find a market niche and acquire a sustainable income. Even in the case of established software houses with mature software products and large customer bases, the problem of increased demand for software updates is significant and difficult to tackle. Therefore, many companies have started to employ lean development approaches, aiming to be competitive and to release products and services as fast as possible.

In this context, software engineering has been a core discipline in the contemporary business model. A well-thought software engineering process that allows rapid software development and easy software change/evolution is critical for achieving focus on product market-fit and user acceptance. And, from all software engineering-related concepts, the one with the most major impact on the process is that of software reuse. Reuse is about effectively integrating existing software components or even employing extracted knowledge from them to produce new ones. Reuse brings multiple benefits, including time and/or effort reduction during development, easier maintenance, and higher quality (when reuse is practiced correctly).

This book is our first structured effort toward understanding this challenge of reuse and toward effectively addressing it in a data-driven manner. We argue that mining the vast amount of software engineering data residing in online repositories can greatly assist the software team in the different phases of the software engineering life cycle. The application areas of this book span across the phases of requirements elicitation and specification extraction, software development, and quality assurance.

In the area of requirements elicitation and specification extraction, we aspire to confront the challenge of modeling requirements in multimodal formats. Natural language processing techniques and semantics are employed to parse functional requirements written in natural language, graphical storyboards, and/or UML diagrams, and subsequently extract and store specifications into ontologies. The aim is not limited to modeling, but also includes subsequently mining the extracted specifications to provide recommendations for requirements.

Our second area of application, software development, poses several interesting reuse challenges, which we separate according to their level of code granularity. Initially, the book proposes a set of tools, including a code search engine as well as a test-driven reuse system, for finding reusable components. These tools are based on information retrieval techniques and employ syntax-aware mining to ensure that the retrieved components address the needs of the developer. At a more fine-grained level, the integration of components and, more generally, the integration of external libraries/frameworks are studied from the scope of the well-known snippet mining problem. Our methods in this case accept natural language queries and employ clustering and scoring mechanisms to extract the most useful snippets for each scenario.

The third area studied in this book involves evaluating the quality of the retrieved components, and especially their potential for reuse, i.e., their reusability. The initial contribution in this area is a recommendation system that evaluates components both from a functional and from a reusability perspective. Consequently, we propose an enhanced reusability assessment model of source code components based on their preference/reuse rate by the developer community.

All in all, this book aspires to provide a unified solution, a set of mining methodologies that can be used in conjunction with contemporary software engineering to produce better software. And this solution is exercised through practicing reuse at all phases of software development. In the ever changing field of software engineering, we believe that this book will at least cultivate a discussion toward creating better software with minimal cost and effort, which has the potential to affect important economic and social aspects of everyday life.

Thessaloniki, Greece   Themistoklis Diamantopoulos
January 2020   Andreas L. Symeonidis

Contents

Part I Introduction and Background

1 Introduction 3
 1.1 Overview 3
 1.2 Scope and Purpose of the Book........................... 5
 1.3 Progress and Contribution of the Book................... 6
   1.3.1 Requirements Mining 7
   1.3.2 Source Code Mining............................. 7
   1.3.3 Quality Assessment 8
 1.4 Overview of the Chapters 9
 1.5 Supplementary Material................................ 11
 References 12

2 Theoretical Background and State-of-the-Art 13
 2.1 Software Engineering 13
 2.2 Software Reuse.. 14
   2.2.1 Overview of Reuse and Reuse Landscape 14
   2.2.2 Software Engineering Data 16
   2.2.3 Component Reuse Challenges 21
 2.3 Software Quality...................................... 22
   2.3.1 Overview of Quality and Its Characteristics 22
   2.3.2 Quality Metrics 24
   2.3.3 Software Testing 28
 2.4 Analyzing Software Engineering Data 29
   2.4.1 Data Mining................................... 29
   2.4.2 Data Mining on Software Engineering Data 32
   2.4.3 Overview of Reuse Potential and Contributions 36
   2.4.4 An Integrated View of Software Reuse 38
 References 39

Part II   Requirements Mining

3   Modeling Software Requirements 47
    3.1   Overview 47
    3.2   State of the Art on Requirements Elicitation
          and Specification Extraction 48
    3.3   From Requirements to Specifications 51
          3.3.1   System Overview 51
          3.3.2   Extracting Artifacts from Software Requirements 52
          3.3.3   A Parser for Automated Requirements Annotation 60
          3.3.4   From Software Artifacts to Specifications 64
    3.4   Case Study 69
    3.5   Conclusion 72
    References 73

4   Mining Software Requirements 75
    4.1   Overview 75
    4.2   State of the Art on Requirements Mining 76
          4.2.1   Functional Requirements Mining 76
          4.2.2   UML Models Mining 79
    4.3   Functional Requirements Mining 81
    4.4   UML Models Mining 85
    4.5   Evaluation 90
          4.5.1   Functional Requirements Mining 90
          4.5.2   UML Models Mining 93
    4.6   Conclusion 94
    References 95

Part III   Source Code Mining

5   Source Code Indexing for Component Reuse 101
    5.1   Overview 101
    5.2   Background on Code Search Engines 102
          5.2.1   Code Search Engines Criteria 102
          5.2.2   Historical Review of Code Search Engines 103
          5.2.3   AGORA Offerings 105
    5.3   The AGORA Code Search Engine 106
          5.3.1   Overview 106
          5.3.2   Analyzers 108
          5.3.3   Document Indexing 110
          5.3.4   Relevance Scoring 114
    5.4   Searching for Code 115
          5.4.1   AGORA Search Features 115
          5.4.2   AGORA Search Scenarios 116
          5.4.3   Example Usage Scenario 122

5.5   Evaluation 125
      5.5.1   Evaluation Mechanism and Dataset 125
      5.5.2   Evaluation Results 127
5.6   Conclusion 131
References 131

6   Mining Source Code for Component Reuse 133
6.1   Overview 133
6.2   Background on Recommendation Systems in Software
      Engineering 134
      6.2.1   Overview 134
      6.2.2   Code Reuse Systems 135
      6.2.3   Test-Driven Reuse Systems 137
      6.2.4   Motivation 139
      6.2.5   Mantissa Offerings 140
6.3   Mantissa: A New Test-Driven Reuse Recommendation
      System 141
      6.3.1   Overview 141
      6.3.2   Parser 143
      6.3.3   Downloader 143
      6.3.4   Miner 146
      6.3.5   Transformer 152
      6.3.6   Tester 154
      6.3.7   Flow Extractor 156
6.4   Mantissa User Interface and Search Scenario 156
      6.4.1   Mantissa User Interface 156
      6.4.2   Example Search Scenario 158
6.5   Evaluation 160
      6.5.1   Evaluation Framework 161
      6.5.2   Comparing Mantissa with CSEs 163
      6.5.3   Comparing Mantissa with FAST 167
      6.5.4   Comparing Mantissa with Code Conjurer 168
6.6   Conclusion 171
References 172

7   Mining Source Code for Snippet Reuse 175
7.1   Overview 175
7.2   State of the Art on Snippet and API Mining 176
7.3   CodeCatch Snippet Recommender 179
      7.3.1   Downloader 179
      7.3.2   Parser 180
      7.3.3   Reusability Evaluator 181

7.3.4   Readability Evaluator 181
7.3.5   Clusterer 182
7.3.6   Presenter 184
7.4   Example Usage Scenario............................... 185
7.5   Evaluation 187
7.5.1   Evaluation Framework 187
7.5.2   Evaluation Results 189
7.6   Conclusion 190
References 191

8   Mining Solutions for Extended Snippet Reuse 193
8.1   Overview 193
8.2   Data Collection and Preprocessing 194
8.2.1   Extracting Data from Questions.................. 194
8.2.2   Storing and Indexing Data 195
8.3   Methodology 197
8.3.1   Text Matching 197
8.3.2   Tag Matching................................ 198
8.3.3   Snippet Matching 198
8.4   Example Usage Scenario............................... 198
8.5   Evaluation 200
8.6   Conclusion 202
References 202

Part IV   Quality Assessment

9   Providing Reusability-Aware Recommendations 207
9.1   Overview 207
9.2   QualBoa: Reusability-Aware Component
Recommendations................................... 208
9.2.1   High-Level Overview of QualBoa 208
9.2.2   Downloading Source Code and Metrics 209
9.2.3   Mining Source Code Components 210
9.2.4   Recommending Quality Code 212
9.3   Evaluation 214
9.4   Conclusion 216
References 216

10   Assessing the Reusability of Source Code Components 219
10.1   Overview 219
10.2   Background on Reusability Assessment 220
10.3   Correlating Reusability with Popularity 221
10.3.1   GitHub Popularity as Reusability Indicator 221
10.3.2   Modeling GitHub Popularity 222

10.4 Reusability Modeling 223
    10.4.1 Benchmark Dataset......................... 223
    10.4.2 Evaluation of Metrics' Influence on Reusability 223
10.5 Reusability Estimation............................ 228
    10.5.1 System Overview 228
    10.5.2 Model Construction 228
10.6 Evaluation 231
    10.6.1 Reusability Estimation Evaluation 231
    10.6.2 Example Reusability Estimation 233
10.7 Conclusion................................... 234
References 234

Part V Conclusion and Future Work

11 Conclusion 239

12 Future Work................................... 241

List of Figures

Fig. 2.1   Typical activities in software engineering 14
Fig. 2.2   The reuse landscape 16
Fig. 2.3   Software engineering data 17
Fig. 2.4   ISO/IEC 25010:2011 quality characteristics
       and their sub-characteristics 23
Fig. 2.5   Data mining tasks and indicative techniques 31
Fig. 2.6   Data mining for software engineering: goals,
       input data used, and employed mining techniques 33
Fig. 2.7   Overview of data mining for software reuse and book
       contributions 36
Fig. 3.1   Example scenario using the language of project
       ReDSeeDS [4] 48
Fig. 3.2   Example scenario and code skeleton using JBehave [7] 49
Fig. 3.3   Example relationship between classes using
       CM-Builder [14] 50
Fig. 3.4   Overview of the conceptual architecture of our system 52
Fig. 3.5   Static ontology of software projects 54
Fig. 3.6   Properties of the static ontology 54
Fig. 3.7   Screenshot of the requirements editor 55
Fig. 3.8   Dynamic ontology of software projects 57
Fig. 3.9   Properties of the dynamic ontology 58
Fig. 3.10   Screenshot of the storyboard creator 59
Fig. 3.11   Pipeline architecture for parsing software requirements 60
Fig. 3.12   Example annotated instance using the hierarchical
       annotation scheme 61
Fig. 3.13   Annotated requirements of project Restmarks 62
Fig. 3.14   Aggregated ontology of software projects 65
Fig. 3.15   Properties of the aggregated ontology 66
Fig. 3.16   Schema of the YAML representation 68
Fig. 3.17   Excerpt of the annotated functional requirements of project
       restmarks 70

Fig. 3.18   Storyboard diagram "Add bookmark" of project restmarks.   71
Fig. 3.19   Example YAML file for project restmarks   72
Fig. 4.1   Cluster editor screen of DARE [19]   77
Fig. 4.2   Example clustering of requirements for a smart home
            system [18].   78
Fig. 4.3   Data model of the UML differencing technique of Kelter
            et al. [26]   80
Fig. 4.4   Example annotated requirement   81
Fig. 4.5   Example WordNet fragment where record is the most
            specific class of account and profile   82
Fig. 4.6   Example use case diagram for project Restmarks   85
Fig. 4.7   Example use case diagram for matching with the
            one of Fig. 4.6   87
Fig. 4.8   Example activity diagram for project Restmarks.   88
Fig. 4.9   Example activity diagram for matching with the one
            of Fig. 4.8.   89
Fig. 4.10   Example depicting **a** the functional requirements of Restmarks,
            b the recommended requirements, and **c** the recommended
            requirements visualized.   91
Fig. 4.11   Visualization of recommended requirements including
            the percentage of the correctly recommended requirements
            given support and confidence   92
Fig. 4.12   Classification results of our approach and the approach
            of Kelter et al. for **a** use case diagrams and **b** activity
            diagrams.   93
Fig. 5.1   The architecture of AGORA.   107
Fig. 5.2   The regular expression of the CamelCase analyzer.   109
Fig. 5.3   List of Java stop words.   110
Fig. 5.4   Home page of AGORA   115
Fig. 5.5   Example query for a class "stack" with two methods,
            one named "push" with return type "void" and one named
            "pop" with return type "int".   117
Fig. 5.6   Example response for the query of Fig. 5.5   117
Fig. 5.7   Example query for a class with name containing "Export",
            which extends a class named "WizardPage" and the Java
            file has at least one import with the word "eclipse" in it   118
Fig. 5.8   Example response for the query of Fig. 5.7   119
Fig. 5.9   Example query for project with files extending classes
            "Model", "View", and "Controller" and also files extending
            the "JFrame" class   120
Fig. 5.10   Example response for the query of Fig. 5.9   120
Fig. 5.11   Example query for a snippet that can be used
            to read XML files.   121
Fig. 5.12   Example response for the query of Fig. 5.11   121

Fig. 5.13   Query for a file reader on the Advanced Search page
            of AGORA.. 122
Fig. 5.14   Query for an edit distance method on the Advanced
            Search page of AGORA....................................... 123
Fig. 5.15   Query for a pair class on the Advanced Search page
            of AGORA.. 124
Fig. 5.16   Example sequence of queries for GitHub................... 126
Fig. 5.17   Example modified test case for the result of a code
            search engine 127
Fig. 5.18   Evaluation diagram depicting the average number of
            compilable and tested results for the three code search
            engines... 129
Fig. 5.19   Evaluation diagrams depicting **a** the search length for finding
            compilable results and **b** the search length for finding results
            that pass the tests, for the three code search engines 130
Fig. 6.1   Architecture of PARSEWeb [9] 137
Fig. 6.2   Example screen of Code Conjurer for a loan calculator
            component [13]... 138
Fig. 6.3   Example screen of S6 for a converter of numbers
            to roman numerals [35]..................................... 139
Fig. 6.4   The architecture/data flow of Mantissa................... 142
Fig. 6.5   Example signature for a class "Stack" with methods
            "push" and "pop".. 142
Fig. 6.6   Example AST for the source code example of Fig. 6.5 143
Fig. 6.7   Example sequence of queries for Searchcode.............. 144
Fig. 6.8   Example sequence of queries for GitHub.................. 145
Fig. 6.9   Regular expression for extracting method declarations
            from snippets 146
Fig. 6.10   Algorithm that computes the similarity between two sets
            of elements, U and V, and returns the best possible
            matching as a set of pairs *MatchedPairs* 148
Fig. 6.11   String preprocessing steps............................... 149
Fig. 6.12   Example signature for a "Stack" component with two methods
            for adding and deleting elements from the stack............. 151
Fig. 6.13   Example vector space representation for the query "Stack"
            and the retrieved result "MyStack"........................ 151
Fig. 6.14   Example signature for a class "Customer" with methods
            "setAddress" and "getAddress"............................ 153
Fig. 6.15   Example retrieved result for component "Customer" 153
Fig. 6.16   Example modified result for the file of Fig. 6.15 154
Fig. 6.17   Example test case for the "Customer" component
            of Fig. 6.14.. 155
Fig. 6.18   Screenshot depicting the search page of Mantissa............ 157
Fig. 6.19   Screenshot depicting the results page of Mantissa 158

Fig. 6.20   Example signature for a component that reads
            and writes files   159
Fig. 6.21   Example write method of a component that reads
            and writes files   159
Fig. 6.22   Code flow for the write method of a component
            that handles files.   160
Fig. 6.23   Example signature for a component that finds the MD5
            of a string.   160
Fig. 6.24   Example test case for the "MD5" component of Fig. 6.23   161
Fig. 6.25   Example signature for a serializable set component   161
Fig. 6.26   Example test case for the serializable set component
            of Fig. 6.25.   162
Fig. 6.27   Example for constructing a set of MD5 strings
            for the files of folder folderPath and writing
            it to a file with filename setFilename   162
Fig. 6.28   Evaluation diagram depicting the average number
            of compilable and tested results for the three code search
            engines and the two Mantissa implementations.   165
Fig. 6.29   Evaluation diagram depicting the average search length
            for finding results that pass the tests for the three
            code search engines and the two Mantissa implementations   167
Fig. 6.30   Evaluation diagram depicting the average number
            of compilable and tested results for FAST and for
            the two Mantissa implementations   169
Fig. 7.1   Screenshot of UP-Miner [2]   177
Fig. 7.2   Example screenshot of eXoaDocs [6].   178
Fig. 7.3   CodeCatch system overview.   179
Fig. 7.4   Example snippet for "How to read a CSV file"   180
Fig. 7.5   Example snippet for "How to read a CSV file" using
            Scanner.   182
Fig. 7.6   Silhouette score of different number of clusters for the example
            query "How to read a CSV file".   184
Fig. 7.7   Silhouette score of each cluster when grouping into
            5 clusters for the example query "How to read a CSV file"   184
Fig. 7.8   Screenshot of CodeCatch for query "How to read
            a CSV file", depicting the top three clusters.   185
Fig. 7.9   Screenshot of CodeCatch for query "How to read
            a CSV file", depicting an example snippet   186
Fig. 7.10   Example snippet for "How to upload file to FTP" using
            Apache Commons.   187
Fig. 7.11   Reciprocal Rank of CodeCatch and Google for the three
            most popular implementations (I1, I2, I3) of each query   190
Fig. 8.1   Example stack overflow question post   195
Fig. 8.2   Example sequence for the snippet of Fig. 8.1   197

Fig. 8.3   Example sequence extracted from a source code snippet 199
Fig. 8.4   Example code snippet from stack overflow post about
            "how to delete file from ftp server using Java?" 200
Fig. 8.5   Percentage of relevant results in the first 20 results
            of each query, evaluated for questions with snippet
            sequence length larger than or equal to 1 201
Fig. 8.6   Percentage of relevant results in the first 20 results
            of each query, evaluated for questions with snippet
            sequence length larger than or equal to **a** 3 and **b** 5 202
Fig. 9.1   The architecture of QualBoa 208
Fig. 9.2   Example signature for class "Stack" with methods
            "push" and "pop" 208
Fig. 9.3   Boa query for components and metrics 210
Fig. 9.4   Evaluation results of QualBoa depicting **a** the average
            precision and **b** the average reusability score
            for each query 215
Fig. 10.1   Diagram depicting the number of stars versus the number
            of forks for the 100 most popular GitHub repositories 222
Fig. 10.2   Package-level distribution of API Documentation
            for all repositories 225
Fig. 10.3   Package-level distribution of API Documentation
            for two repositories 226
Fig. 10.4   API Documentation versus star-based reusability score 227
Fig. 10.5   Overview of the reusability evaluation system 228
Fig. 10.6   Average NRMSE for all three machine learning algorithms 229
Fig. 10.7   Reusability score distribution at **a** class and **b** package
            level 230
Fig. 10.8   Boxplots depicting reusability distributions for three
            human-generated (■) and two auto-generated (■) projects,
            a at class level and **b** at package level (color online) 232

List of Tables

Table 2.1   Source code hosting services   18
Table 2.2   C&K Metrics [35]   25
Table 3.1   Example instantiated OWL classes for the storyboard
            of Fig. 3.10   59
Table 3.2   Ontology instances for the entities of Restmarks   63
Table 3.3   Properties for the ontology instances of the FR4
            of Restmarks   63
Table 3.4   Properties of the aggregated ontology   65
Table 3.5   Classes mapping from static and dynamic ontologies
            to aggregated ontology.   67
Table 3.6   Properties mapping from static and dynamic ontologies
            to aggregated ontology.   67
Table 3.7   Instantiated classes Resource, Property,
            and Activity for restmarks   71
Table 4.1   Sample association rules extracted from the dataset   83
Table 4.2   Activated rule heuristics for a software project.   84
Table 4.3   Matching between the diagrams of Figs. 4.6 and 4.7   87
Table 4.4   Matching between the diagrams of Figs. 4.8 and 4.9   89
Table 4.5   Evaluation results for the recommended requirements   92
Table 5.1   Popular source code search engines   104
Table 5.2   Feature comparison of popular source code search
            engines and AGORA   105
Table 5.3   The mapping of projects documents of the AGORA
            index   111
Table 5.4   The mapping of files documents of the AGORA index   111
Table 5.5   The inner mapping of a class for the files documents
            of AGORA.   113
Table 5.6   Evaluation dataset for code search engines.   126

Table 5.7   Compiled results and results that passed the tests for the three
             code search engines, for each query and on average, where
             the format for each value is number of tested results/number
             of compiled results/total number of results   128
Table 5.8   Search length for the compiled and passed results
             for the three code search engines, for each result
             and on average, where the format for each value
             is passed/compiled   130
Table 6.1   Feature comparison of popular test-driven reuse systems
             and Mantissa   140
Table 6.2   String preprocessing examples   149
Table 6.3   Dataset for the evaluation of Mantissa against code search
             engines   163
Table 6.4   Compiled and passed results out of the total returned
             results for the three code search engines and the two
             Mantissa implementations, for each result and on average,
             where the format is passed/compiled/total results   164
Table 6.5   Search length for the passed results for the three code
             search engines and the two Mantissa implementations,
             for each result and on average   166
Table 6.6   Dataset for the evaluation of Mantissa against FAST   168
Table 6.7   Compiled and passed results out of the total returned
             results for FAST and the two Mantissa implementations,
             for each result and on average, where the format for each
             value is passed/compiled/total results   169
Table 6.8   Dataset for the evaluation of Mantissa against Code
             Conjurer   170
Table 6.9   Results that passed the tests and response time for the two
             approaches of Code Conjurer (Interface-based and
             Adaptation) and the two implementations of Mantissa
             (using AGORA and using GitHub), for each result
             and on average   171
Table 7.1   API clusters of CodeCatch for query "How to upload
             file to FTP"   186
Table 7.2   Statistics of the queries used as evaluation dataset   188
Table 8.1   Example lookup table for the snippet of Fig. 8.1   196
Table 8.2   Example stack overflow questions that are similar
             to "uploading to FTP using java"   199
Table 9.1   The reusability metrics of QualBoa   209
Table 9.2   The reusability model of QualBoa   213
Table 9.3   Dataset for the evaluation of QualBoa   214
Table 9.4   Evaluation results of QualBoa   215
Table 10.1   Categories of static analysis metrics related to reusability   222

Table 10.2   Overview of static metrics and their applicability
             on different levels.................................... 224
Table 10.3   Dataset statistics.................................... 231
Table 10.4   Metrics for classes and packages with different reusability.... 233

Part I
Introduction and Background

Chapter 1
Introduction

1.1 Overview

Nowadays, software is practically everywhere: information systems for large-scale organizations and companies, web and mobile applications, and embedded systems are some of the types of software systems defined. The software development industry has already identified the need for efficient software development and maintenance in order to produce better software, which in turn has impact on various aspects of everyday life. Software engineering has always been a challenging discipline, mainly due to the fact that it keeps evolving along with the irrefutable evolution of the Computer and Information Science. Current advances, including the evolution of the Internet and the collaborative open-source paradigm, have altered the way we confront challenges in different areas of software engineering, and subsequently the way we develop software with respect to everyday needs.

In order to cope with the challenges posed from these advances, software engineering has also evolved. Although software has been developed for a long time before the dawn of the information age [1], the design methodologies, the employed tools, and even the philosophy around software development have changed significantly during this time. One such fundamental change is the leap from structured programming to object-oriented languages in the 1980s, to functional and scripting languages in the 1990s, and even to metaprogramming for the years that followed. Software engineering process models have also evolved, starting from simple waterfall models and moving on to the iterative, component-based, and later the agile paradigm [2]. All in all, times have changed and so has software; the software industry has grown, new practices have arisen, and even the purpose of developing software has changed.

Nevertheless, the main challenges that traditionally concern the software engineering industry remain the same (or are at least similar). According to the 1994 CHAOS report of the Standish Group [3], in the United States more than \$250 billion were spent each year on IT application development of approximately 175,000 projects. These costs indeed point to a growing industry and may not be on their own

© Springer Nature Switzerland AG 2020

T. Diamantopoulos and A. L. Symeonidis, *Mining Software Engineering Data for Software Reuse*, Advanced Information and Knowledge Processing, https://doi.org/10.1007/978-3-030-30106-4_1

alarming; the staggering fact, however, stems from the conclusions of the report, which estimates that $81 billion out of these company/government funds were spent on (ultimately) canceled software projects, and another $59 billion were spent on challenged projects that may be completed; however, they exceeded their initial cost and time estimates. Even now, more than 20 years later, the software industry faces more or less the same challenges; the updated report of 2015 by the same organization, which is publicly available [4], suggests that 19% of software projects usually fail, while 52% are challenged (cost more than originally estimated and/or delivered late). This means that only 29%, i.e., less than one-third of the projects, are delivered successfully in a cost-effective and timely manner.[1] The latest CHAOS report at the time of the writing was released in late 2018 [5] and still indicates that only 36% of projects succeed, i.e., almost 2 out of 3 projects fail or surpass the original estimates.

The reasons for these unsettling statistics are diverse. A recent study [6] identified 26 such reasons, including the inadequate identification of software requirements, the inaccurate time/cost estimations, the badly defined specifications by the development team, the lack of quality checks, the unfamiliarity of the team with the involved technologies, etc. Although avoiding some of these reasons is feasible under certain circumstances, the real challenge is to manage all or at least most of them, while keeping the cost and effort required to a minimum. And this can only be accomplished by taking the proper decisions at all stages of the software development life cycle.

Decision-making, however, is not trivial; it is a challenging process that requires proper analysis of all relevant data. The set of methodologies used to support decision-making in the area of software engineering are commonly described by the term *software intelligence*.[2] In specific, as noted by Hassan and Xie [8], software intelligence "offers software practitioners up-to-date and pertinent information to support their daily decision-making processes". Software practitioners are not only developers, but anyone actively participating in the software life cycle, i.e., product owners, software maintainers, managers, and even end users. The vision of software intelligence is to provide the means to optimize the software development processes, to facilitate all software engineering phases, including requirements elicitation, code writing, quality assurance, etc., to allow for unlocking the potential of software teams, and generally to support decision-making based on the analysis of data.

As already mentioned, though, the main notion of using data for decision-making has been pursued for quite some time. Today, however, there is a large potential in

---

[1] https://www.infoq.com/articles/standish-chaos-2015.

[2] Software intelligence can be actually seen as a specialization of *business intelligence*, which, according to Gartner analyst Howard Dresner [7], is an umbrella term used to describe "concepts and methods to improve business decision-making by using fact-based support systems". Although this contemporary definition can be traced in the late 1980s, the term business intelligence is actually older than a century and is first attributed to Richard M. Devens, who used it (in his book "Cyclopaedia of Commercial and Business Anecdotes: Comprising Interesting Reminiscences and Facts, Remarkable Traits and Humors ... of Merchants, Traders, Bankers ... etc. in All Ages and Countries", published by D. Appleton & Company in 1868) to describe the ability of the banker Sir Henry Furnese to gain profit by taking successful decisions according to information received by his environment.

this aspect, because today more than ever the data are available. The introduction of new collaborative development paradigms and the open-source software initiatives, such as the Open Source Initiative,[3] have led to various online services with different scopes, including code hosting services, question-answering communities, bug tracking systems, etc. The data residing in these services can be harnessed to build intelligent systems that can in turn be employed to support decision-making for software engineering.

The answer to how these data sources can be effectively parsed and utilized is actually a problem of *reuse*. Software reuse is a term referred for the first time by the Bell engineer Douglas McIlroy, who suggested using already existing components when developing new products [9]. More generally, software reuse can be defined as "the use of existing assets in some form within the software product development process" [10], where the assets can be software engineering artifacts (e.g., requirements, source code, etc.) or knowledge derived from data (e.g., software metrics, semantic information, etc.) [11]. Obviously, reuse is already practiced to a certain degree, e.g., software libraries are a good example of reusing existing functionality. However, the challenge is to fully unlock the potential of the data in order to truly take advantage of the existing assets and subsequently improve the practices of the software industry. Thus, one may safely argue that the data is there, and the challenge becomes to analyze them in order to extract the required knowledge and reform that knowledge in order to successfully integrate it in the software development process [12].

1.2 Scope and Purpose of the Book

The data of a software project comprise its source code as well as all files accompanying it, i.e., documentation, readme files, etc. In the case of using some type of version control, the data may also include information about commits, release cycles, etc., while integrating also a bug tracking system may offer issues, bugs, links to commits that resolve these, etc. In certain cases, there may also be access to the requirements of the project, which may come in various forms, e.g., text, UML diagrams, etc. Finally, at a slightly higher level, the raw data of these sources can serve to produce processed data in the form of software metrics [13], used to measure various quality aspects of the software engineering process (e.g., quality metrics, people metrics, etc.). So, there is no doubt that the data are there; the challenge is to utilize them effectively toward achieving certain goals.

The ambition posed by this book is to employ these software engineering data in order to facilitate the actions and decisions of the various stakeholders involved in a software project. In specific, one could summarize the scope of the book in the following fragment:

---

[3]http://opensource.org/.

Apply data mining techniques on software engineering data in order to facilitate the different phases of the software development life cycle

This purpose is obviously broad enough to cover different types of data and different results; however, the aim is not to exhaust all possible methodologies in the area of software engineering. The focus is instead on producing a unified approach and illustrating that data analysis can potentially improve the software process.

We view the whole exercise under the prism of *software reuse* and illustrate how already existing data can be harnessed to assist in the development process. The proposed solution advances research in the requirements elicitation phase, continues with the phases of design and specification extraction where focus is given on software development and testing, and also contributes in software quality evaluation. Using the products of our research, we aspire that development teams should be able to save considerable time and effort, and navigate more easily through the process of creating software.

1.3 Progress and Contribution of the Book

The area of mining software engineering data is currently on a spree of producing novel ideas and research paths. This research area has grown up to its current form in the beginning of the twenty-first century, when the research community posed interesting questions with respect to software, its development process, and its quality.

The place of this book in this active scientific area is delicate: we do not wish to challenge current research from a results/efficiency perspective; instead, we aspire to produce a unified reuse-oriented approach that would direct the research toward truly aiding the software industry. As such, we focus on some of the most fault-prone phases of software engineering and attempt not only to produce findings beyond the current state of the art, but also to illustrate how combining these findings can be of great use to development teams. In specific, we focus on the following areas:

- Requirements mining,
- Source code mining, and
- Quality assessment.

Although we do not exhaust the potential application areas of mining techniques on software engineering data, we focus on three of the most prominent ones in current research. There are of course several other interesting fields where mining techniques can provide useful insight, such as software design or testing. However, these are not included in the main scope of this book, and they are only discussed with respect to the three main areas defined above (e.g., testing is discussed along with code mining in the context of test-driven development), leaving their more detailed analysis for future research.

Finally, the state of the art in the three aforementioned main areas and the relevant progress of this book are analyzed in the following subsections. Note that here a high-level analysis is provided, while exhaustive literature reviews are performed in the corresponding chapters.

1.3.1 Requirements Mining

The requirements elicitation and the specification extraction phases of a software project are crucial. Inaccurate, incomplete, or undefined requirements seem to be the most common reason of failure [6], as they may have repercussions in all phases of the software development process. In this book, we focus on requirements modeling and mining; as such, our goal is to design a model capable of storing software requirements that will also enable requirements reuse. The current state of the art in this area is mostly focused on instantiating models using domain-specific vocabularies, with the purposes of performing validation, recommending new requirements, and recovering dependencies among requirements or between requirements and software components.

Most current methods can be effective under certain circumstances; however, they are confined to specific domains. Our contribution in this aspect is a novel domain-agnostic model that supports storing software requirements from multimodal input and further adheres to the specifics of the static and the dynamic view of software projects. Our model is flexible enough to cover different requirements elicitation paradigms and is coupled with supporting tools to be instantiated. Finally, we further advance on current requirement recommendation systems, where we design a mining methodology that effectively checks whether the requirements of a software project are complete and recommends new requirements, thus allowing engineers to improve the existing functionality and the data flow/business flow of their project.

1.3.2 Source Code Mining

Source code mining is an area that has lately gained considerable traction, as faster high-quality development usually translates to faster time to market with minimized cost. The idea of directly aiding developers in writing code has initially been covered mostly by the appropriate tools (e.g., Integrated Development Environments—IDEs) and frameworks (languages, libraries, etc.). During the latest years, a clear trend toward a component-based philosophy of software can be observed. As such, developers require ready-to-use solutions for covering the desired software functionality. These solutions can be reusable components of software projects, API functionalities of software libraries, responses to queries in search engines and/or question-answering systems, etc. Thus we focus our work in this aspect on different areas in order to cover a large part of what we call source code or functionality reuse.

Our first contribution is a specialized search engine for source code, commonly called a *Code Search Engine (CSE)*. Several CSEs have been developed mainly during the latest 15 years; however, most of them have more or less suffered from the same issues: they do not offer advanced queries according to source code syntax, they are not integrated with code hosting services and thus can easily become outdated, and they do not offer complete well-documented APIs. By contrast, our CSE performs extensive document indexing, allowing for several different types of queries, while it is also continuously updated from code hosting services. Furthermore, its API allows it to serve as a source of input for the construction of several tools; in fact, we ourselves use it as the core/input for building different services on top.

Possibly, the most straightforward integration scenario for a CSE is a connection with a *Recommendation System in Software Engineering (RSSE)* [14]. Once again, we identify an abundance of such systems, most of them following the same path of actions: receiving a query for a component from the developer, searching for useful components using a search engine, performing some type of scoring to rank the components, and possibly integrating a component to the source code of the developer. To achieve progress beyond the state of the art, we have designed and developed a novel RSSE, which first of all receives queries as component interfaces that are usually clear for the developer. Our RSSE downloads components on demand and ranks them using a syntax-aware mining model, while it also automatically transforms them to allow seamless integration with the source code of the developer. Finally, the results can be further assessed using tests supplied by the developer in order to ensure that the desired functionality is covered.

A similar, yet also useful, contribution to the above is offered by our research in snippet and API mining. In specific, we develop a methodology to improve on the focused retrieval of API snippets for common programming queries. We are based on the source code syntax in order to incorporate information that can be successfully used to provide useful ready-to-use snippets. With respect to current approaches, our progress in this case can be measured quantitatively, as we prove that our methodology can offer snippets of API calls for different queries. As another contribution in this aspect, we also provide a methodology for validating the usefulness and the correctness of snippets using information from question-answering services.

1.3.3 Quality Assessment

Having developed a software component/prototype/product, one may evaluate it along two axes: whether it successfully covers the functionality for which it was designed and developed (usually expressed by functional requirements), and whether it covers certain non-functional criteria [15]. Measuring what we call software quality has long been a peculiar art, as the term "quality" is highly context-dependent and may mean different things to different people [16]. During the latest years, the state of the practice has dictated measuring certain quality characteristics defined by the international ISO standard [17] using software metrics [13]. Thus, research in this area largely focuses on proposing new metrics derived from different software engineering data.

Initially, we focus on the code reuse RSSEs described in the previous subsection. As already noted, several of these systems may be effective from a functional perspective, as they recommend components that are suitable for the query of the developer. However, they do not assess the reusability of the recommended components, i.e., the extent to which a component is reusable. As a result, we design and develop an RSSE that assesses development solutions not only from a functional, but also from a non-functional perspective. Apart from recommending useful components, our system further employs an adaptable model based on static analysis metrics for assessing the reusability of each component.

In an effort to better evaluate the quality and the reusability of software components, we also conduct further research on the area of quality and reusability assessment using static analysis metrics. There is an abundance of approaches concerning the construction of quality evaluation models, i.e., models that receive as input values of source code metrics and output a quality score based on whether the metrics exceed certain thresholds. These approaches, however, usually rely on expert knowledge either for specifying metric thresholds directly or for providing a set of ground truth quality values required for automatically determining those thresholds. Our contribution in this aspect is a methodology for determining the quality of a software component as perceived by developers, which refrains from expert representations of quality; instead, we employ popularity as expressed by online services and prove that a highly reused (or favorited) component is usually of high quality.

Given our findings, we design and develop a system that uses crowdsourcing information as ground truth and train models capable of estimating the reusability degree both of a software project as a whole and individually for different components. Our system focuses on the main quality attributes and properties (e.g., complexity, coupling, etc.) that should be assessed in order to determine whether a software component is reusable.

1.4 Overview of the Chapters

As already mentioned in Sects. 1.2 and 1.3 of this chapter, the scope of this book lies in a reusability-oriented perspective of software which tackles three different software phases. As a result, the text is hopefully a comfortable and compact read from top to bottom. However, as each software engineering phase on its own may be of interest to certain readers, extra effort has been made to split the text into self-contained parts, which therefore allow one to focus on a specific set of contributions, as these were outlined in Sect. 1.3. Thus, after the first part of this book, one could skip to any of the three subsequent parts and finish with the conclusions of the last part. The book consists of five distinct parts, while each part includes a set of different chapters. The overview of the book is provided as follows:

Part I: Introduction and Background
This part contains all introductory information of the book, explaining its scope and the relevant scientific areas. It includes the following chapters:

Chapter 1. Introduction
The current chapter that included all information required to understand the scope and purpose of this book, as well as a high-level overview of the contributions of this book.

Chapter 2. Theoretical Background and State of the art
This chapter contains all the background knowledge required to proceed with the rest of the parts and chapters of the book. All necessary definitions about software engineering, software reuse, software quality, and any other relevant areas are laid out and discussed.

Part II: Requirements Mining
This part contains the contributions of the book that are relevant to the area of requirements engineering. It includes the following chapters:

Chapter 3. Modeling Software Requirements
This chapter presents our model for software requirements that supports instantiation from multimodal formats and allows performing different actions on the stored requirements. As such, our model is the basis of the following chapter.

Chapter 4. Mining Software Requirements
Our mining methodologies for functional requirements and UML models is presented in this chapter. The methodologies are explained step by step, and useful examples of requirements recommendations are provided in each case.

Part III: Source Code Mining
This part contains the contributions of the book that are relevant to the area of software development. It includes the following chapters:

Chapter 5. Source Code Indexing for Component Reuse
This chapter presents the design of a CSE that is oriented toward source code reuse. Our system facilitates reuse in component level, snippet level, and project level and can be useful for answering multiple research questions.

Chapter 6. Mining Source Code for Component Reuse
This chapter initially presents the design of an RSSE that is oriented toward test-driven reuse. We explain how our system can be useful for locating reusable components and evaluate it on typical datasets against similar systems.

Chapter 7. Mining Source Code for Snippet Reuse
This chapter presents our methodology for API snippet mining. We

design a system for discovering different API implementations as answers to common programming questions.

Chapter 8. Mining Solutions for Extended Snippet Reuse

This chapter presents our methodology for validating one's source code snippets using information from question-answering systems. We design a system for easily locating relevant question posts for common queries.

Part IV: Quality Assessment

This part contains the contributions of the book that are relevant to the area of software quality. It includes the following chapters:

Chapter 9.   Providing Reusability-aware Recommendations

This chapter presents an RSSE that assesses software components both from a functional and from a reusability perspective. Thus, the system comprises a syntax-aware matching mechanism and a model that assesses the reusability of a component.

Chapter 10.   Assessing the Reusability of Source Code Components

This chapter presents a novel way of assessing the reusability of source code components by employing popularity and reuse information as ground truth.

Part V: Conclusion and Future Work

This part contains the conclusive remarks of the book as well as ideas for potential future work. It includes the following chapters:

Chapter 11. Conclusion

The conclusions of the book are presented in this chapter, including any remarks for the lessons learned and the contributions offered during the course of this book.

Chapter 12. Future Work

This chapter includes ideas for potential extensions toward all areas of contributions of this book.

1.5   Supplementary Material

An important aspect of any scientific work is the potential to reproduce it and to improve it. As such, for this book, we would like to provide the resources required for reproducing our findings, which shall hopefully help other researchers make progress on the relevant scientific areas. However, providing resources via multiple links is a precarious practice; links to running prototypes and websites tend to become deprecated, and the data are then unavailable. Instead, we decided to create a website that comprises all the different resources that are relevant to this book, including datasets, models, tools (prototypes and/or source code), etc. The website is available at:

https://thdiaman.github.io/mining-software-reuse-book/

References

1. Robert Curley (2011) Architects of the Information Age. Britannica Educational Publishing
2. Sommerville I (2010) Software Engineering, 9th edn. Addison-Wesley, Harlow, England
3. Standish Group (1994) The CHAOS Report (1994). Technical report, Standish Group
4. Standish Group (2015) The CHAOS Report (2015). Technical report, Standish Group
5. Standish Group (2018) The CHAOS Report (2018): Decision Latency Theory: It's All About the Interval. Technical report, Standish Group
6. Montequín VR, Fernández SC, Fernández FO, Balsera JV (2016) Analysis of the success factors and failure causes in projects: comparison of the spanish information and communication technology ICT sector. Int J Inf Technol Proj Manag 7(1):18–31
7. Power DJ (2007) A Brief History of Decision Support Systems. http://dssresources.com/history/dsshistory.html. Retrieved November, 2017
8. Hassan AE, Xie T (2010) Software intelligence: the future of mining software engineering data. In: Proceedings of the FSE/SDP Workshop on Future of Software Engineering Research, FoSER '10. ACM, New York, NY, USA, pp 161–166
9. McIlroy MD (1968) Components Mass-produced Software. In: Naur P, Randell B (eds) Software Engineering; Report of a Conference sponsored by the NATO Science Committee. NATO Scientific Affairs Division, Brussels, Belgium, NATO Scientific Affairs Division. Belgium, Brussels, pp 138–155
10. Lombard Hill Group (2017) What is Software Reuse? http://www.lombardhill.com/what_reuse.htm. Retrieved November, 2017
11. Frakes William B, Kang Kyo (2005) Software reuse research: status and future. IEEE Trans Softw Eng 31(7):529–536
12. Krueger CW (1992) Software Reuse. ACM Comput Surv 24(2):131–183
13. Fenton N, Bieman J (2014) Software Metrics: A Rigorous and Practical Approach, 3rd edn. CRC Press Inc, Boca Raton, FL, USA
14. Robillard M, Walker R, Zimmermann T (2010) Recommendation systems for software engineering. IEEE Softw 27(4):80–86
15. Pressman Roger, Maxim Bruce (2019) Software Engineering: A Practitioner's Approach, 9th edn. McGraw-Hill Inc, New York, NY, USA
16. Pfleeger SL, Kitchenham B (1996) Software quality: The elusive target. IEEE Software. pp 12–21
17. ISO/IEC 25010:2011 (2011) https://www.iso.org/standard/35733.html. Retrieved: November 2017

Chapter 2
Theoretical Background and State-of-the-Art

2.1 Software Engineering

Ever since the dawn of software engineering, which is placed toward the late 1960s,[1] it was clear that a structured way of creating software was eminent; individual approaches to development were simply unable to scale to proper software systems. As a result, the field has been continuously expanding with the introduction of different methodologies for developing software. And although there are several different methodologies (e.g., waterfall, iterative, agile), all designed to create better software, the main activities of the software development process are common to all of them.

An analytical view of the typical activities/stages of software engineering, which is consistent with current literature [1, 3, 4], can be found in Fig. 2.1. These activities are more or less present in all software systems. For the development of any project, one initially has to identify its requirements, translate them to specifications, and subsequently design its architecture. After that, the project is implemented and tested, so that it can be deployed and sustained by the required operations. Even after its completion, the project has to be maintained in order to continue to be usable. Note that non-maintainable projects tend to become easily deprecated, for a number of reasons, such as the lack of security updates that may make the system vulnerable to certain threats.

Thus, we focus on the different activities defined in Fig. 2.1 (e.g., requirements elicitation, source code implementation, etc.), and on the connections between subsequent activities (e.g., translating requirements to specifications).

---

[1] According to Sommerville [1], the term was first proposed in 1968 at a conference held by NATO. Nowadays, we may use the definition by IEEE [2], which describes software engineering as "the application of a systematic, disciplined, quantifiable approach to the development, operation, and maintenance of software; that is, the application of engineering to software".

© Springer Nature Switzerland AG 2020
T. Diamantopoulos and A. L. Symeonidis, *Mining Software Engineering Data for Software Reuse*, Advanced Information and Knowledge Processing,
https://doi.org/10.1007/978-3-030-30106-4_2

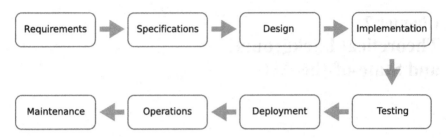

Fig. 2.1 Typical activities in software engineering

Arguably, one can minimize the risks of developing software in a number of ways. In this book, we focus on two such risk mitigation approaches, that of (indexing and) reusing existing solutions and that of continuously assessing the selected processes, the software engineering activities, and their outcome. These two ways correspond roughly to the areas of *software reuse* and *software quality assessment*, which are analyzed in the following sections.

2.2 Software Reuse

2.2.1 Overview of Reuse and Reuse Landscape

Software reuse[2] is a notion first introduced in the late 1960s, and, since then, has dictated a new philosophy of developing software. In this section, we shall first focus on the benefits of reuse and then indicate possible reuse sources and areas of application.

In rather simplistic terms, reuse means using already existing knowledge and data instead of reinventing the wheel. Obviously, this bears several advantages [6]. The first and possibly the most important benefit of reuse is that of time and effort reduction. When reusing existing components, instead of building software from scratch, a development team can produce software faster and the end product will then have a better time to market. The developers themselves will have more time on their hands to dedicate on other aspects of the project, which are also influenced by reuse, such as the documentation. The second important gain from software reuse is that of ensuring the final product is of high quality. Development can be quite hard, as creating and thoroughly testing a component may be almost impossible within certain time frames. Reusing an existing component, however, provides a way of

---

[2]In the context of this book, software reuse can be defined as "the systematic application of existing software artifacts during the process of building a new software system, or the physical incorporation of existing software artifacts in a new software system" [5]. Note that the term "artifact" corresponds to any piece of software engineering-related data and covers not only source code components, but also software requirements, documentation, etc.

separating between one's software and the component at hand, as the production and maintenance of the latter is practically outsourced. Even if the component is not third party, it is always preferred to maintain a set of different clearly separated components, rather than a large complex software system. With that said, the reuse of certain non-properly constructed components may raise risks in the quality of the end product; however, in this section, we focus on proper reuse methodologies and discuss the concept of whether components are suitable for reuse in the next section.

As we dive in deeper to the concept of reuse, a question that arises is what are the methodologies of reuse, or, in other words, what is considered reuse and what not. An interesting point of view is provided by Krueger [7] in this aspect, who maintains that reuse is practiced everywhere, from source code compilers and generators to libraries and software design templates. The point, however, that Krueger made more than 20 years ago is that we have to consider reuse as a standard software engineering practice. Today, we indeed have accomplished to develop certain methodologies toward this direction. As Sommerville interestingly comments [1], informal reuse actually takes place regardless of the selected software development process.[3] The methodologies that have been developed practically facilitate the incorporation of software reuse with software engineering as a whole. They can be said to form a so-called *reuse landscape* [1], which is shown in Fig. 2.2. This landscape involves options to consider both prior and during the development of new software. It includes methodologies such as component-based software engineering, reusable patterns (design or architectural) for constructing new software, and existing functionality wrapped in application frameworks or program libraries.

Another more recent viewpoint on reuse is given by Capilla et al. [8]. The authors note that reuse has been initially practiced in a domain-specific basis. Domain analysis was used to understand the software artifacts and processes, and produce reusable information for software engineering [9–11]. In this context, reuse has also been the target of several Model-Driven Engineering (MDE) approaches that transform existing components to make them compatible with different frameworks or even languages [12]. However, the major evolution of the reuse landscape has its traces in component reuse. As Capilla et al. [8] note, component-based software engineering reuse has its roots in the development of object-oriented languages and frameworks that allowed for components with increased modularity and interoperability.

Finally, one cannot dismiss the importance of services and open data toward forming the current reuse landscape [8, 13]. Deploying reusable components as services

---

[3]A formalized alternative is that of reuse-oriented software engineering, which can be defined as defining requirements, determining whether some can be covered by reusing existing components, and finally designing and developing the system with reuse in mind (i.e., including designing component inputs/outputs and integrating them). However, this is a development style already followed by several development teams as part of their process, regardless of its traditional or modern nature. As a result, we argue that there is no need to force development teams to use a different methodology and do not focus on the reuse-oriented perspective as a formalized way of developing software. Instead, we present here a set of methodologies and indicate the potential from using software engineering data in order to actually incorporate reuse as a philosophy of writing software.

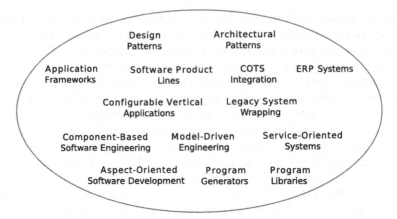

Fig. 2.2 The reuse landscape

(or, lately, microservices) has allowed for off-the-shelf reuse, while achieving not only high modularity, but also improved scalability. Open data, on the other hand, have provided multiple opportunities for building applications with increased inter-operability in certain domains (e.g., transportation). Albeit not in the narrow scope of this book, open data are today abundant[4] and can provide testbeds for several interesting research questions about software reuse.

2.2.2 Software Engineering Data

In the context of this book, we plan to facilitate reuse when applied on different phases of the software development processes. We focus on reuse from a data-oriented perspective and argue that one can at any time exercise reuse based on the data that are available to harness. Data in software engineering come from various sources [14]. An indicative view of data that can be classified as software engineering-related is shown in Fig. 2.3, where the major sources are shown in bold. Note also that the format for the data of Fig. 2.3 can be quite diverse. For example, requirements may be given in free text, as structured user stories, as UML diagrams, etc. Also, for each project, different pieces of data may or may not be present, e.g., there may be projects without bug tracking systems or without mailing lists. Another important remark concerns the nature of the data: some types of data, such as the source code or the requirements, may be available directly from the software project, whereas others, such as the quality metrics, may be the result of some process. In any case, these information sources are more or less what developers must be able to comprehend,

---

[4]There are several open data repositories, one of the most prominent being the EU Open Data Portal (https://data.europa.eu/euodp/en/home), which contains more than 14000 public datasets provided by different publishers.

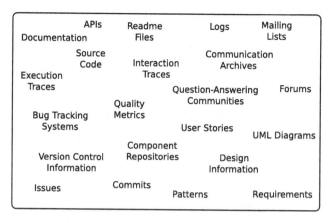

Fig. 2.3 Software engineering data

utilize, and take advantage of, in order to practice the reuse paradigm and, in a way, improve on their development process and their product.[5]

At first glance, all the above sources may seem hard to gather. Moreover, even though a development team can maintain a database of these data for its own projects, enriching it also with third-party data, which admittedly in software reuse can be very helpful, is hard. However, in modern software engineering, the necessary tools for collecting these data are available, while the data themselves are abundant in online repositories. As a result, the field of software engineering has changed, as developers nowadays use the web for all challenges that may come out during the development process [16]. A recent survey has shown that almost one-fifth of the time of developers is spent on trying to find answers to their problems online [17]. Further analyzing the sources where developers look for their answers, Li et al. [18] found out that there is a rather large variety of them, including technical blogs, question-answering communities, forums, code hosting sites, API specifications, code example sites, etc.

Several services have been developed in order to accommodate this new way of creating software, and we are today in a position where we can safely say that "social coding" is the new trend when it comes to developing software. An indicative set of services that have been developed to support this new paradigm are analyzed in the following paragraphs.

---

[5]Note that we focus here mainly on software engineering data that are available online. There are also other resources worth mining, which form what we may call the *landscape of information* that a developer is faced with when joining a project, a term introduced in [15]. In specific, apart from the source code of the project, the developer has to familiarize himself/herself with the architecture of the project, the software process followed, the dependent APIs, the development environment, etc.

Table 2.1 Source code hosting services

Service	URL	# Users	# Projects	Year
SourceForge	https:// sourceforge.net/	3,700,000	500,000	1999
GNU Savannah	https://savannah. gnu.org/	90,000	4,000	2001
Launchpad	https://launchpad. net/	4,000,000	40,000	2004
GitHub	https://github. com/	31,000,000	100,000,000	2008
Bitbucket	https://bitbucket. org/	5,000,000	–	2008
GitLab	https://gitlab. com/	100,000	550,000	2011

Note that these numbers are approximate. Also, these statistics are valid for the time of the writing; however, they are subject to change quite often. The interested reader is referred to the online article https://en.wikipedia.org/wiki/Comparison_of_source_code_hosting_facilities, where complete and updated statistical information about code hosting services can be found

2.2.2.1   Code Hosting Services

Arguably, the most important type of services is that of code hosting services, as they are the ones that allow developers to host their source code online. As such, they offer the required raw source code information, which is usually linked to by online communities. An indicative set of popular source code hosting services is shown in Table 2.1.

As we may see in this table, the peak for the establishment of code hosting services can be placed around 2008, when GitHub and Bitbucket joined the already large SourceForge and the three together nowadays host the projects of more than 30 million developers. If we consider the statistics for the total number of projects in these services, we can easily claim that there are currently millions of projects hosted online in some source code hosting facility.

2.2.2.2   Question-Answering Sites

Several online communities have been lately established so that developers can communicate their potential problems and find specific solutions. Even when developers do not use directly these services, it is common to be redirected to them from search engines as their queries, and subsequently their problems may match those already posted by other developers. The most popular of these services is Stack Overflow,[6]

---

[6]https://stackoverflow.com/.

a collective programmer community[7] created in 2008. With more than 11 million subscribed users, and over 50 million developers accessing the service every month, Stack Overflow practically offers a huge pool of solutions for different problems related to programming. Its parent site, Stack Exchange, today offers several different question-answering services,[8] with several of them dedicated to the software engineering process. Indicatively, there are sites for software engineering,[9] for project management,[10] for software quality assurance and testing,[11] etc.

There are also several other sites similar to Stack Exchange, such as Quora[12] and Reddit.[13] Quora was established in 2010 as a generic question-answering community. Given, however, that it supports tags for specific programming problems, it is often chosen by programmers when searching for answers to their questions. Reddit, on the other hand, which was established as early as 2005, comprises a set of communities, several of which are relevant to software development, such as the programming community[14] with more than 2 million subscribers, the software development community[15] with approximately 37 thousand subscribers, and the software architecture community[16] with more than 8 thousand subscribers. Finally, in this category of services, we may also add the forums of multi-purpose programming websites, such as CodeProject,[17] which was established in 1999 as a resource hub, including programming tutorials, articles, newsletters, etc. The discussion boards of the service have more than 10 million members that can take part or navigate in several different boards for programming languages, frameworks, etc.

2.2.2.3 Software Project Directories

Up until now, we have outlined services that allow developers to index their source code and find online possible solutions to their problems. Another quite important type of services that is widely used today is that of software project indexes. The term is rather broad, covering different types of software; however, the common point among all is that they support the component-based development paradigm by allowing the developer to reuse certain components in an intuitive and semi-automated

---

[7]The term is attributed to Jeff Atwood, the one of the two creators of Stack Overflow, the other being Joel Spolsky [19, 20].

[8]https://stackexchange.com/sites.

[9]https://softwareengineering.stackexchange.com/.

[10]https://pm.stackexchange.com/.

[11]https://sqa.stackexchange.com/.

[12]https://www.quora.com/.

[13]https://www.reddit.com/.

[14]https://www.reddit.com/r/programming/.

[15]https://www.reddit.com/r/softwaredevelopment/.

[16]https://www.reddit.com/r/softwarearchitecture/.

[17]https://www.codeproject.com/.

manner. Typical services in this category include the maven central repository,[18] the Node Package Manager (npm) registry,[19] and the Python Package Index (PyPI).[20]

All these services comprise an online index of software libraries as well as a tool for retrieving them at will. Maven is used primarily for Java projects and has indexed at the time of the writing more than 300.000 unique artifacts.[21] The developer can browse or search the index for useful libraries, which can then be easily integrated by declaring them as dependencies in a configuration file. Similar processes are followed by the npm registry. The index hosts more than 1 million packages for the JavaScript programming language, allowing developers to easily download and integrate what they need in their source code. The corresponding repository for the Python programming language, PyPI, contains more than 120.000 Python libraries, which can be installed in one's distribution using the pip package management system.

Software directories are, however, not limited to libraries. There are, for instance, directories for web services, such as the index of ProgrammableWeb,[22] which contains the APIs of more than 22.000 web services. Developers can navigate through this directory to find a web service that can be useful in their case and integrate it using its API.

2.2.2.4 Other Sources

There is no doubt that the sources of data discussed above comprise huge information pools. Most of them contain raw data, i.e., source code, forum posts, software libraries, etc. A different category of systems is that of systems hosting software *metadata*. These include data from bug tracking systems, continuous integration facilities, etc. Concerning bug tracking systems, one of the most popular ones is the integrated issues service of GitHub. Other than that, we may also note that there are several online services that facilitate the management of software development. An example such service is Bugzilla,[23] which functions as an issue/bug tracking system. The service is used by several open-source projects, including the Mozilla Firefox browser,[24] the Eclipse IDE,[25] and even the Linux Kernel,[26] and the hosted information is available to anyone for posting/resolving issues, or for any other use that may possibly be relevant to one's own project (e.g., when developing a plugin for Eclipse or Firefox).

---

[18] http://central.sonatype.org/.

[19] https://www.npmjs.com/.

[20] https://pypi.python.org/pypi.

[21] https://search.maven.org/stats.

[22] https://www.programmableweb.com/.

[23] https://www.bugzilla.org/.

[24] https://bugzilla.mozilla.org/.

[25] https://bugs.eclipse.org/bugs/.

[26] https://bugzilla.kernel.org/.

As for other sources of metadata, one can refer to online indexes that have produced data based on source code, on information from version control systems, etc. Such services are the Boa Language and Infrastructure[27] or the GHTorrent project,[28] which index GitHub projects and further extract information, such as the Abstract Syntax Tree of the source code. The scope of these services, however, is mostly addressed in other sections of this book, as they are usually research products, which operate upon some form of primal unprocessed data (as in this case Boa employs the data of GitHub).

2.2.3 Component Reuse Challenges

One can easily note that the amount of software engineering data that are at any time publicly available online is huge. This new data-oriented software engineering paradigm means that developers can find anything they need, from tools to support the management and/or the requirements specification and design phases, to reusable components, test methodologies, etc. As a result, the main challenge reside on (a) finding an appropriate artifact (either tool or component or generally any data-driven solution), (b) understanding its semantics, and (c) integrating it into one's project (source code or even process).

When these steps are followed effectively, the benefits in software development time and effort (and subsequently cost) are actually impressive. The problem, however, is that there are various difficulties arising when developers attempt to exercise this type of reuse in practice. The first and probably most important difficulty is that of localizing what the developer needs; given the vastness of the information available online, the developer can easily be intimidated and/or not be able to localize the optimal solution in its case. Furthermore, even when the required components are found and retrieved, there may be several issues with understanding how they function, e.g., certain software library APIs may not be sufficiently documented. This may also make the integration of a solution quite difficult. Finally, this paradigm of search and reuse can easily lead to quality issues, as the problems of third-party components may propagate to one's own product. Even if a component is properly designed, there is still some risk involved as poor integration may also lead to quality decline. Thus, as software quality is an important aspect of reuse, its background is analyzed in the following section, while the overall problem described here is again discussed with all relevant background in mind in Sect. 2.4.

---

[27] http://boa.cs.iastate.edu/.

[28] http://ghtorrent.org/.

2.3 Software Quality

2.3.1 Overview of Quality and Its Characteristics

Software quality is one of the most important areas of software engineering, yet also one of the most hard to define. As David Garvin effectively states [21], "quality is a complex and multifaceted concept". Although quality is not a new concept, the effort to define it is somewhat notorious[29] and its meaning has always been different for different people [3]. According to the work of Garvin, which was later adopted by several researchers [23, 24], the quality of a product is relevant to its performance, to whether its features are desirable, to whether it is reliable, to its conformance to certain standards, to its durability and serviceability, and to its aesthetics and perception from the customer. Another early definition of the factors that influence software quality is that of McCall et al. [25–27]. The authors focus on aspects relevant to (a) the operation of the product, measured by the characteristics of correctness, reliability, usability, integrity, and efficiency; (b) its ability to change, measured by the characteristics of maintainability, flexibility, and testability; and (c) its adaptiveness, measured by the characteristics of portability, reusability, and interoperability. Though useful as definitions, these characteristics can be hard to measure [3].

Nowadays, the most widely accepted categorization of software quality is that of the latest quality ISO, a standard developed to identify the key characteristics of quality, which can be further split into sub-characteristics that are to some extent measurable. The ISO/IEC 9126 [28], which was issued in 1991, included the quality characteristics of functionality, reliability, usability, efficiency, maintainability, and portability. An interesting note for this standard is that these characteristics were further split into sub-characteristics, which were in turn divided into *attributes* that, in theory at least, are actually measurable/verifiable. Of course, the hard part, for which no commonly accepted solution is provided until today, is to define the metrics used to measure each quality attribute that also are in most cases product- and technology-specific. The most major update in the standard was the ISO/IEC 25010 [29], which was first introduced in 2011 and comprises eight product quality characteristics that are depicted in Fig. 2.4, along with their sub-characteristics.

Each of the proposed ISO characteristics refers to different aspects of the software product. These aspects are the following:

- Functional suitability: the extent to which the product meets the posed requirements;
- Performance efficiency: the performance of the product with respect to the available resources;
- Portability: the capability of a product to be transferred to different hardware or software environments;

---

[29]Indicatively, we may refer to an excerpt of the rather abstract "definition" provided by Robert Persig [22]: "Quality... you know what it is, yet you don't know what it is".

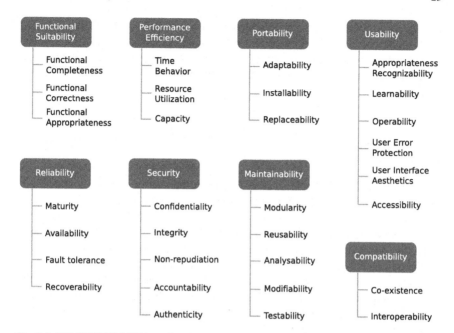

Fig. 2.4 ISO/IEC 25010:2011 quality characteristics and their sub-characteristics

- Compatibility: the degree to which the product can exchange information with other products;
- Usability: the degree to which the product can be used effectively and satisfactorily by the users;
- Reliability: the efficiency with which the product performs its functions under specified conditions and time;
- Security: the degree to which the data of the product are protected and accessed only when required by authorized parties; and
- Maintainability: the effectiveness and efficiency with which the product can be modified, either for corrections or for adaptions to new environments or requirements.

As already noted, there are several difficulties when trying to measure the quality of a software product. Hence, there is an ongoing effort in designing *metrics* that would effectively correspond to quality attributes, which, as already mentioned, can be mapped to the sub-characteristics and the characteristics of Fig. 2.4. Several metrics have been defined in the past [30, 31], which can be classified according to the scope of the quality management technique, or in other words according to the type of data that is measured. According to the SWEBOK [32],[30] there are four categories of software quality management techniques: static analysis, dynamic analysis,

---

[30]The SWEBOK, short for Software Engineering Body of Knowledge, is an international ISO standard that has been created by the cooperative efforts of software engineering professionals and

testing, and software process measurement techniques. Each of these categories has its own purpose, as defined by current research and practice, while different metrics have been proposed over time as software methodologies change. We analyze these quality management techniques and certain indicative metrics in the following subsections.

2.3.2   Quality Metrics

As already mentioned, there are several types of metrics in software engineering, measuring different aspects of the system under development and the development process itself [30, 31]. If we wish to make an indicative (yet not exhaustive) mapping to the quality characteristics of Fig. 2.4, we can say that characteristics such as maintainability or usability are usually measured by some form of static analysis metrics. On the other hand, dynamic analysis metrics are typically used for measuring characteristics such as reliability or performance efficiency. Testing, which is practically its own discipline so it is analyzed in the following subsection, is mainly used for assessing the functional suitability of the product. Finally, there are also metrics that focus on the software process itself.

2.3.2.1   Static Analysis Metrics

The first category of metrics that we are going to analyze refers to metrics derived by performing some type of analysis on the source code of an application, without executing it. The introduction of this category of metrics can be traced back in the 1970s, with *size* and *complexity* being the two most important properties measured at the time. Widely applied static analysis metrics were the Lines of Code (LoC), the McCabe cyclomatic complexity [33], and the Halstead complexity measures [34]. Though rather simplistic, the LoC metric has been quite useful at the time of its introduction as in certain commonly used languages, like FORTRAN, each non-empty line represented a single specific command. Today, the metric is still useful, however, mostly to have an order of magnitude for source code components. The cyclomatic complexity was defined in 1976 as an effort to measure how complex a software system is. The metric is computed by the number of linearly independent paths through the source code of the software. The set of metrics that was developed in 1977 by Maurice Halstead [34] had also as purpose to measure complexity; however, it was based on the counts of source code operators and operands.

is published by the IEEE Computer Society. The standard aspires to summarize the basic knowledge of software engineering and include reference lists for its different concepts.

Table 2.2 C&K Metrics [35]

ID	Name
WMC	Weighted methods per class
DIT	Depth of inheritance tree
NOC	Number of children
CBO	Coupling between objects
RFC	Response for a class
LCOM	Lack of cohesion in methods

Although the aforementioned metrics are more than 40 years old, they are still used widely (and in most cases correctly[31]) as indicators of program size and complexity. However, the introduction of the object-oriented programming paradigm and especially its establishment in the 1990s has raised new measurement needs, as the source code quality of an object-oriented program is further relevant to the properties of *coupling* and *inheritance*. Hence, several new metrics have been proposed for covering these properties. Indicatively, we may refer to the C&K metrics [35] that are named after their creators, Chidamber and Kemerer, and are shown in Table 2.2.

WMC is usually computed as the sum of the complexities of all methods included in the class under analysis. DIT and NOC refer to the number of superclasses and subclasses of the class, respectively. RFC and CBO are used to model relationships between classes; RFC is typically defined as the number of methods in the class and all external methods that are called by them, while CBO refers to the number of classes that the class under analysis depends on, i.e., their methods/variables that are accessed by the class. The inverse metric also exists and is usually referred to as Coupling Between Objects Inverse (CBOI). Finally, LCOM is a metric for the degree of cohesion between the methods of a class and is related to the number of method pairs with common variables among them, having subtracted the number of method pairs without common variables among them.[32]

Another quite popular set of object-oriented metrics proposed in 1994 [37] is the MOOD metrics suite, short for "Metrics for Object-Oriented Design". The authors constructed metrics for modeling encapsulation, i.e., the Method Hiding Factor (MHF) and the Attribute Hiding Factor (AHF), for modeling inheritance, i.e., Method Inheritance Factor (MIF) and Attribute Inheritance Factor (AIF), as well as

---

[31]A rather important note is that any metric should be used when it fits the context and when its value indeed is meaningful. A popular quote, often attributed to Microsoft co-founder and former chairman Bill Gates, denotes that "Measuring programming progress by lines of code is like measuring aircraft building progress by weight".

[32]This original definition of LCOM has been considered rather hard to compute, as its maximum value depends on the number of method pairs. Another computation formula was proposed later by Henderson-Sellers [36], usually referred as LCOM3 (and the original metric then referred as LCOM1), which defines LCOM as $(M - A/V)/(M - 1)$, where M is the number of methods and A is the number of accesses for the V class variables.

for modeling polymorphism and coupling, i.e., the Polymorphism Factor (POF) and the Coupling Factor (COF), respectively. The metric proposals, however, do not end here; there are several other popular efforts including the suite proposed by Lorenz and Kidd [38]. Nevertheless, the main rationale for all of these metrics is similar and they all measure the main quality properties of object-oriented design using as input the source code of a project.[33]

2.3.2.2   Dynamic Analysis Metrics

As opposed to static analysis, dynamic analysis is performed by executing a program and measuring its behavior. As such, the comparison between the two analyses is quite interesting. Static analysis metrics are usually easier to obtain and are usually available at early stages of software development. On the other hand, dynamic analysis allows also measuring the behavioral aspects of the software, which is not feasible by using static analysis metrics [40]. An indicative set of metrics for this category are the Export Object Coupling (EOC) and Import Object Coupling (IOC) metrics, proposed by Yacoub et al. [41]. The EOC and IOC metrics refer to the number of messages sent or received, respectively, from one object to another. There are several other proposals in current literature, such as the one by Arisholm et al. [42], who designed a set of twelve metrics related to coupling, such as the number of external methods invoked from a class and the number of classes used by a class, all measured at runtime.[34]

Another important axis involves the potential of dynamic analysis to measure characteristics such as reliability.[35] The metrics of reliability are external, as they measure mainly the behavior of a system or component when executing it, without focusing on its inner elements. The metrics of this category and their corresponding ranges are rather broad, as they may be dependent on the specifics of different types of systems [43]. The most popular metrics in this category are the Mean Time Between Failures (MTBF) that measures the amount of failures for a specified execution time, the Mean Time To Failure (MTTF) that refers to the expected time until the first failure of the product, and the Mean Time To Repair (MTTR) that is measured as the expected time for repairing a module after it has failed, a metric that is also highly relevant to hardware systems. Other important metrics include the Probability Of Failure On Demand (POFOD) that measures the probability of a system failing when a request is made, the Rate Of oCcurrence Of Failures (ROCOF) that reflects

---

[33]The interested reader is further referred to [39] for a comprehensive review of object-oriented static analysis metrics.

[34]There are several other metrics in this field; for an outline of popular metrics, the reader is referred to [40].

[35]On the other hand, static analysis is usually focused on the characteristics of maintainability and usability.

the frequency of failures,[36] and the Availability (AVAIL) that refers to the probability that the system is available for use at a given time.

2.3.2.3 Software Process Analysis Metrics

As with the two previous categories of metrics, models of software process metrics cover a wide range of possible applications. This category measures the dynamics of the process followed by a software development team. Some of the first notions in this field are attributed to the IBM researcher Allan Albrecht. In 1979, Albrecht attempted to measure application development productivity by using a metric that he invented, known as the *Function Point (FP)* [44]. As opposed to size-oriented metrics discussed earlier in this section, the FP, which is determined by an empirical relationship based on the domain and the complexity of the software, gave rise to what we call *function-oriented metrics*.[37] If one manages to define FP, then it can be used as a normalization value to assess the performance of the software development process. Out of the several metrics that have been proposed, we indicatively refer to the number of errors (or defects) per FP, the cost per FP, the amount of documentation per FP, the number of FPs per person-month of work, etc.

Lately, the increased popularity of version control systems has also made available another set of metrics, the so-called "change metrics", which constitute a type of metrics that are preferred also for multiple purposes, such as predicting defects or indicating bad decisions in the development process. An indicative set of change metrics was proposed by Moser et al. [46] that defined the number of revisions of a file, the code churn (number of added lines minus number of deleted lines) computed per revision or in total, the change set (number of files committed together to the repository) computed per revision or in total, etc. These metrics can be computed using information from the commits and the issues (bugs) present in version control systems, while their main scope is to predict defects in software projects. The rationale for the aforementioned metrics is quite intuitive; for instance, files (components) with many changes from many people are expected to be more prone to bugs. The change set metrics are also quite interesting, as they indicate how many files are affected in each revision. Building on the intuition that changes affecting multiple files can more easily lead to bugs than single-file commits, Hassan further proposed the entropy of code changes [47], computed for a time interval as $\sum_{k=1}^{n} p_k \cdot log_2 p_k$, where p_k is the probability that the kth file has changed.

As a final note, the metrics analyzed in the previous paragraphs can be applied for several reasons, such as to assess the reusability of components, to detect bugs in source code, and to determine which developer is more suitable for each part of the source code. It is important, however, to understand that the metrics themselves are

---

[36]MTTF is actually the reciprocal of ROCOF.

[37]In fact, the original publication by Albrecht [44] indicated that function-oriented and size-oriented metrics were highly correlated; however, the rise of different programming languages and models quickly proved that their differences are quite substantial [45].

not adequate when trying to assess software quality or possibly to detect defects. That is because, on its own, each metric does not describe a component from a holistic perspective. For instance, having a class with large size does not necessarily mean complex code that is hard to maintain. Equivalently, a rather complex class with several connections to other classes may be totally acceptable if it is rather small in size and well-designed. As a result, the challenge with these metrics usually relies on defining a model [48], i.e., a set of thresholds that, when applied together, properly estimate software quality, detect defects, etc.

2.3.3 Software Testing

According to Sommerville [1], the testing process has two goals: (a) to demonstrate that the software meets the posed requirements and (b) to discover errors in the software. These goals are also well correlated with the view that testing belongs in a wider spectrum, namely, that of verification and validation [3]. If we use the definitions of Boehm [49], verification ensures that we are "building the product right", while validation ensures that we are "building the right product". We shall prefer this interpretation of testing, which is widely accepted,[38] as it allows us to relate it not only to the final product, but also to the requirements of the software.

When testing a software product, one has to take several decisions, two of the most important being what to test and in what amount of detail should each individual test case and subsequently the enclosing test suite go into. Given that testing may require significant time and effort [3], the challenge of optimizing one's resources toward designing a proper testing strategy is crucial. An effective strategy (that of course depends on each particular case) may comprise test cases in four different levels: unit, integration, system, and acceptance testing [52]. Unit testing is concerned with individual components, i.e., classes or methods, and is focused on assessing their offered functionality. Integration testing is used to assess whether the components are also well assembled together and is focused mainly on component interfaces. System testing assesses the integration of system components at the highest level and is focused on defects that arise at this level. Finally, acceptance testing evaluates the software from the point of view of the customer, indicating whether the desired functionality is correctly implemented. Thus, depending on the components of the product at hand, one can select to design different test cases at different granularity.

An even more radical approach is to design the test cases even before constructing the software itself. This approach to writing software is known as *Test-Driven Devel-*

---

[38]There are of course other more focused interpretations, such as the one provided in [50], which defines testing as "the process of executing a program with the intent of finding errors". As, however, times change, testing has grown to be a major process that can add actual value to the software, and thus it is not limited to error discovery. An interesting view on this matter is provided by Beizer [51], who claims that different development teams have different levels of maturity concerning the purpose of testing, with others defining it as equivalent to that of debugging, others viewing it as a way to show that the software works, others using it to reduce product quality risks, etc.

opment (TDD), and although it was introduced with relation to extreme programming and agile concepts [53, 54], it is also fully applicable on traditional software process models. TDD is generally considered as one of the most beneficial software practices, as it has multiple advantages [55]. At first, writing the test cases before writing the code implicitly forces the developer to clarify the requirements and determine the desired functionality for each piece of code. Along the same axis, connecting test cases with requirements can further help in the early discovery of problems in requirements, thus avoiding the propagation of these problems at later stages. Finally, TDD can improve the overall quality of source code as defects are generally detected earlier.

2.4 Analyzing Software Engineering Data

This introduction, though somewhat high level, covers the main elements, or rather building blocks, of software development and serves two purposes: (a) to explain the methodologies followed to build reusable high-quality software products and (b) to identify the data that can be harnessed to further support the software development life cycle. In this section, the aim is to restate the challenge posed by this book and, given the background, discuss how it can be addressed while also making significant progress beyond the current state of the art.

Up until now, we have explained what are the areas of software development that can be improved, and we have also indicated the potential that is derived from the vastness of available software engineering data. The challenge that we now wish to address to is that of analyzing these data to practically improve the different phases of the software development life cycle. The relevant research area that was established to address this challenge is widely known as "Data Mining for Software Engineering". We have already discussed software engineering, so in the following subsection we will focus on data mining, while in Sect. 2.4.2, we will explain how these fields are combined so that data mining methods can assist the software development process. Finally, in Sect. 2.4.3, we outline the potential of employing data mining to support reuse and indicate our contributions in this aspect.

2.4.1 Data Mining

The concept of analyzing data is not new; people have always tried to find some logic in the available data and take decisions accordingly, regardless of the relevant practical or research field. Lately, however, with the rise of the information age during the second half of the previous century, we have put ourselves in a challenging position where the gap between the generation of new data and our understanding of it is constantly growing [56]. As a result, the effective analysis of data has become more eminent than ever. In principle, understanding and effectively using data to extract knowledge is the result of a process called *data analysis*. However, data analysis is mainly acquainted with scientific fields such as statistics, which can of course prove valuable under certain scenarios, yet their applicability is usually limited. In specific, as noted by Tan et al. [57], traditional data analysis techniques cannot easily handle heterogeneous and complex data, they do not scale to large amounts of data (or even data with high dimensionality), and, most importantly, they are confined to the traditional cycle of constructing and (dis)proving hypotheses manually, which is not adequate for analyzing real data and discovering knowledge.

Thus, what differentiates data mining from traditional data analysis techniques is its usage of algorithms from other fields to support knowledge extraction from data. The interdisciplinary nature of the field inclines us to follow the definition proposed by Tan et al. [57]:

> Data mining is a technology that blends traditional data analysis methods with sophisticated algorithms for processing large volumes of data.

This definition is adopted because it encompasses the *sophisticated* nature of the mining techniques, an aspect usually highlighted by several authors.[39] And this sophistication is acquired by mixing the traditional data analysis methods with techniques from the fields of Artificial Intelligence (AI), machine learning, and pattern recognition. As a result, the field as a whole is well spread over statistical methods, such as sampling and hypothesis testing, and AI methods, including, among others, learning theories [59], pattern recognition [60], information retrieval [61], and even evolutionary computing [62].

Given the above discussion, data mining may seem too broad a field; however, its main principles can be summarized by the tasks usually performed, which also provide a rather effective categorization of the employed techniques. A generally accepted perspective [57, 63] is to categorize the tasks of data mining into two categories: descriptive modeling tasks and predictive modeling tasks. The first category refers to extracting patterns from data in order to summarize the underlying relationships, while the objective of the tasks of the second category is to predict the value of certain data attributes based on knowledge extracted from the data. In certain contexts, predictive modeling can actually be seen as an extension of descriptive modeling (although this is not always the case), as one has to first understand the data and subsequently request for certain predictions. An extension to this line of thought is the lately introduced category of prescriptive modeling, which can be seen as the combination of the other two modeling categories, as it is not limited to predictions but rather recommends actions and decisions based on the origin of the predictions; so, it is considered quite useful in business intelligence-related decision-making. Viewed from the prism of Business Analytics [64], descriptive modeling replies to the question of what is going on, predictive modeling replies to the questions of what is going to happen and why, and prescriptive modeling replies to the questions of what should we do and why.

---

[39] An alternative definition provided by Hand et al. [58] involves the notions of "finding unsuspected relationships" and "summarizing data in novel ways". These phrases indicate the real challenges that differentiate data mining from traditional data analysis.

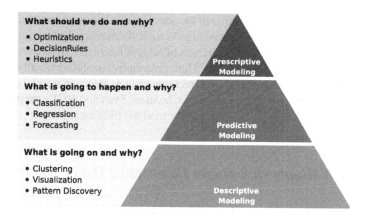

What should we do and why?
- Optimization
- DecisionRules
- Heuristics

Prescriptive Modeling

What is going to happen and why?
- Classification
- Regression
- Forecasting

Predictive Modeling

What is going on and why?
- Clustering
- Visualization
- Pattern Discovery

Descriptive Modeling

Fig. 2.5 Data mining tasks and indicative techniques

These categories of data mining tasks are also shown in Fig. 2.5 along with several indicative types of techniques for each category. The techniques shown in the left part of Fig. 2.5 are only a fraction of those that can be useful for data mining. Although we do not aspire (and do not find it relevant) to analyze all of these techniques here, we will at least provide an abstract description for certain interesting methods, and point to further reading where possible. At first, concerning descriptive modeling, the main point is to try to extract knowledge from the data. As such, an initial thought would be to visualize the data, which is quite useful, yet no always as trivial a task as it may seem. One has to take into account the dimensionality of the data and produce visualizations that depict useful information, while possibly discarding redundant and possibly meaningless information. Thus, another quite important challenge in descriptive modeling is to find out what is worth showing to the analyst. This is accomplished by *pattern discovery*, which is considered as one of the most crucial aspects of data mining [56]. In this context, we may also place unsupervised machine learning methods, such as clustering [65], which is a technique that can organize the data into groups and thus again provide intuition as to (possibly hidden) relations between them. Finally, several modeling approaches can also be used for descriptive knowledge extraction, including methods for fitting distributions to data, mixture models, etc. [60].

Distribution fitting is more often used as part of a predictive modeling task and is relevant to the supervised regression algorithms. Similar to regression, which can be defined as the process of fitting a model to certain data, one can also define forecasting, with the second, however, implying that the data are sequential [60]. Another quite large category of techniques that are relevant to this task are the classification techniques. In both regression (and forecasting) and classification, the initial goal is to construct a model based on existing data, so that certain decisions can be made for new data points. The output of regression, however, lies in a continuous range of values, whereas in a classification problem the new data points need to be classified in a class out of a predefined set of classes [66]. Finally, prescriptive modeling includes techniques similar to the ones already discussed, however, involving further the element of decision-making. As such, the machine learning algorithms selected in this case may have as output sets of rules or heuristics. Moreover, they are mainly focused on model optimization. Thus, prescriptive modeling usually involves comprehensive classification/regression techniques (e.g., decision trees), rule-based decision support systems (e.g., Learning Classifier Systems) [67], computational intelligence algorithms (e.g., fuzzy inference models) [62], etc.

2.4.2 Data Mining on Software Engineering Data

Upon having a basic understanding of the fields of software engineering and data mining, the next step is to combine these two fields to produce added value with respect to the software engineering process. This subsection includes a definition of this broad research area, indicating its different axes, the current limitations, and certain amount of discussion on points that have been considered for improvement in this book. This area involves the application of data mining techniques on software engineering data, with the purpose of extracting knowledge that can improve the software engineering process. This definition practically means that there are countless applications on this area. According to Xie et al. [68], we can distinguish research work along three axes: the goal, the input data used, and the employed mining technique. This decomposition is better visualized by Monperrus [69] as in Fig. 2.6.

As one may notice, Fig. 2.6 involves almost everything that was discussed so far in this chapter. It involves the different sources and types of software engineering data, the employed algorithms that lie in the interdisciplinary field of data mining, and of course a set of tasks that are all part of the software development process. To further analyze the field, we have chosen to follow a categorization similar to that in [70], which distinguishes the work on this research area according to the software engineering activity that is addressed each time.[40] From a rather high-level view, we may focus on the activities of requirements elicitation and specification extraction, software development, and testing and quality assurance. The current state of research in these categories is analyzed in the following paragraphs in order to better define the research field as a whole and indicate the focus of this book. We view each research area separately and illustrate what one can do (or cannot do) today by using the appropriate data along with the methods proposed (or methods not already proposed or methods to be proposed by us in the following chapters). Note, however, that the following paragraphs do not serve as exhaustive reviews of

---

[40] Another possible distinction is that of the type of the input data used, which, according to Xie et al. [68], fall into three large categories: sequences (e.g., execution traces), graphs (e.g., call graphs), and text (e.g., documentation).

Goal
- find and fix bugs
- find optimizations
- improve code quality
- improve documentation
- assist developers *(e.g. completion)*
- manage people
- configure systems
- estimate costs
- ...

Input Data
- source code
- binary code
- configuration files
- documentation
- execution traces
- communication *(mailing-lists,forums)*
- version control data *(commits, etc.)*
- ...

Mining Technique
- classification *(e.g. nearest neighbors)*
- regression
- clustering
- structure mining *(e.g. frequent itemsets)*
- graph mining
- summarization
- metaheuristics *(e.g. genetic algorithm)*
- ...

Fig. 2.6 Data mining for software engineering: goals, input data used, and employed mining techniques

the state of the art in each of the discussed areas. Instead, they provide what we may call the current research practice; for each area, we explain its purpose and describe a representative fraction of its current research directions, while we focus on how the underlying challenges should be addressed. For an extensive review of the state of the art in each of these areas, the reader is referred to each corresponding chapter of this book.

2.4.2.1   Requirements Elicitation and Specification Extraction

Requirements elicitation can be seen as the first step of the software development process; knowing what to build is actually necessary before starting to build it. The research work in this area is focused on optimizing the process of identifying requirements and subsequently on extracting specifications. As such, it usually involves the process of creating a model for storing/indexing requirements as well as a methodology for mining them for various purposes. Given a sufficient set of requirements and an effective model that would capture their structure and semantics (through careful annotation), one could employ mining techniques to perform validation [71, 72], recommend new requirements [11, 73, 74], detect dependencies among requirements [75], etc. If we further add data from other sources, e.g., stakeholder preference, software components, documentation, etc., then additional interesting tasks can be accomplished, including distributing of requirements according to stakehold-ers [75], or even the lately popular challenge of recovering dependencies between requirements and software components [76], known as requirements tracing [70].

Although the aforementioned ideas are quite interesting, they are not usually explored adequately for two reasons. The first is that requirements are usually pro-vided in formats that are hard to mine (e.g., free text, diverse diagram types), while the second is that there do not actually exist many annotated requirements datasets in current literature. As a result, most state-of-the-art approaches are restricted to domain-specific models and cannot incorporate higher order semantics. In this book, we construct a novel domain-agnostic model that allows instantiation from multiple formats of requirements and facilitates employing mining techniques for certain of the applications outlined above.

2.4.2.2   Software Development

The development of a product is seemingly the phase that has the most added value; therefore, it is important that it is performed correctly. The mining methodologies in this phase are numerous, span along quite diverse directions, and have several appli-cations.[41] The input data are mainly in the form of source code and possibly project-related information (e.g., documentation, readme files), while the inner models used for mining involve various formats, such as Abstract Syntax Trees (ASTs), Control Flow Graphs (CFGs), and Program Dependency Graphs (PDGs). As we move toward a reuse-oriented component-based software development paradigm, one of the most popular challenges in this research area is that of finding reusable solutions online. Thus, there are several research works on designing Code Search Engines (CSEs) [78–80] and their more specialized counterparts, Recommendation Systems in Soft-ware Engineering (RSSEs) [81–83], that can be used for locating and retrieving reusable components. These approaches perform some type of syntax-aware min-ing on source code and rank the retrieved components according to certain criteria, so that the developer can select the one that is most useful in his/her case. Finally, there are also systems designed with the test-driven development paradigm in mind, which further test the retrieved components to confirm that they meet the provided requirements [84–86] and which are usually referred to as Test-Driven Reuse (TDR) systems [87].

Although these systems may be effective when examined under the prism of generic software reuse, they also have several limitations. Most CSEs, for example, do not offer advanced syntax-aware queries and are not always integrated with code hosting services, and thus their potential to retrieve useful components is limited by the size of their index. As for RSSEs, their pitfalls are mainly located in the lack of semantics when searching for components, the lack (or inadequacy) of code trans-formation capabilities, and their purely functional one-sided results assessment [88].

---

[41]We analyze here certain popular directions; however, we obviously do not exhaust all of them. An indicative list of applications of source code analysis for the interested reader can be found at [77].

In this book, we design and develop a CSE and two RSSEs, in an effort to overcome these issues. Our systems employ syntax-aware mining and code transformation capabilities and further provide information about the reusability of components, so that the developer can select the optimal one for his/her case and easily integrate it into his/her source code.

Research in this field can also be divided with respect to different levels of granularity. Apart from higher level components, there is a growing interest for finding snippets that may be used as solutions to common programming questions, or snippets that illustrate the usage of an API. In this area, there are again several valuable contributions, which can be distinguished by the task at hand as well as by the source of the data used. There are approaches focused on formulating useful examples for library APIs [89, 90], others discovering useful snippets from question-answering services [91], and others retrieving snippets for common programming queries [92, 93]. The main challenge of all of these approaches is whether they can indeed be effective and provide added value to the developer. As such, in this book, we focus on certain metrics that can quantitatively assess whether the retrieved solutions are correct (e.g., measuring relevance in snippet mining), effective (e.g., how one can easily search in question-answering systems to improve on existing source code), and practically useful (e.g., indicate what is the best way to present results in a snippet recommendation system).

2.4.2.3 Testing and Quality Assurance

The third and final research area analyzed as part of the broad software mining field is that of quality assurance. Given that quality is a process that is relevant to all phases of a software project, the research work also ranges within all these phases, as well as within software evolution [94]. We are going to discuss current approaches with respect to the static analysis described in Sect. 2.3.2. Although static analysis can be used for different purposes, such as assessing the quality of software components and localizing bugs, in our case we focus on assessing the quality of software components. To better highlight the potential of this challenge, we initially construct a component reuse system that employs a quality model to assess the recommended components not only from a functional but also from a quality perspective.

In this context, the challenge is to construct a model that shall determine a set of thresholds for the given static analysis metrics. The model then can be used for assessing the quality of components [95–97] or for finding potentially defect-prone locations [98, 99], by checking whether their metrics' values exceed those thresholds. In this context, several researchers have involved expert help in order to define acceptable thresholds [100, 101]. However, expert help is usually subjective, context-specific, and may not be always available, thus not allowing to adapt a quality assessment system [102]. To this end, there have been multiple efforts toward automatically defining metrics' thresholds from some ground truth quality values [95–97]. As, however, it is hard to obtain an objective measure of software quality, these approaches also typically resort to expert help. To refrain from the aid of experts, in this book, we use as ground truth the popularity and the reusability of software components, as this is measured by online services, which thus provide an indicator of quality as perceived by developers. Given this ground truth, we construct a quality assessment system that provides a reusability score for software components.

2.4.3 Overview of Reuse Potential and Contributions

The potential of employing data mining techniques to support reuse with respect to the contributions of this book is illustrated in Fig. 2.7. At the top of this figure, we place all information that is available to the developers, which may come from multiple online and offline sources. The data from these sources have been outlined in Sect. 2.2.2 and may include source code, documentation, software requirements, data from version control or quality assurance systems, etc. Though useful, the information from these sources is primary and may not be directly suited for employing mining techniques with the purpose of software reuse. Hence, we may define the so-called knowledge extraction layer, which is placed at the middle of Fig. 2.7. This layer involves intermediate systems that receive the raw primary information and processes it to provide data that are suitable for mining.

In the context of software requirements, the challenge is usually a modeling problem. Given a pool of software requirements that may be expressed in different formats (e.g., user stories, UML diagrams, etc.), the first step is to effectively extract their elements (e.g., extract the main actors or actions), and store them in a model that

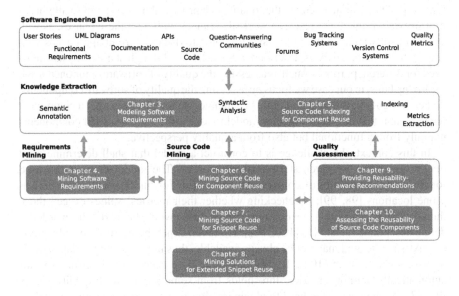

Fig. 2.7 Overview of data mining for software reuse and book contributions

will be suitable for reuse. This is explored in Chap. 3 of this book. Having such a knowledge extraction scheme and model, one can then employ data mining techniques in order to facilitate reuse among different requirements artifacts. We provide two such examples of reuse-oriented mining, one for functional requirements and one for UML models (Chap. 4).

The next area of interest is that of source code reuse. This area can actually be seen as the product of the former one; software requirements practically define the functionality of the system and source code development is used to meet it. This link is also bidirectional as good software should of course comply with the defined requirements, while the requirements should also be up to date and thus conform to the developed software. For an overall exploration of this link, we may direct the interested reader in the broad field of automated software engineering, which explores the connection between software requirements/specifications and source code/application development, while also attempting to automate it as a process. In the context of this book, we discuss this connection in Chap. 3, where we refer to a web service scenario and indicate how requirements and source code can form the discussed bidirectional and traceable link.

The area of source code mining is quite broad, involving the proposal of several tools for knowledge extraction from source code, documentation, etc. The primary information in this case resides in code repositories, version control systems, etc., while the relevant tools may be CSEs or other indexing facilities, analysis tools for extracting syntax-aware information from source code (e.g., API call sequences), etc. Our contribution in this aspect is discussed in Chap. 5, where we design and develop a CSE that extracts syntactic information from source code and facilitates reuse in different levels (component level, snippet level, and project/multi-component level). Given our proposed tool as well as similar tools that are available online (e.g., other CSEs, QA engines, etc.), we are able to produce valuable contributions in different aspects of source code reuse. As such, we construct three systems, one for test-driven reuse at component level (Chap. 6), one for recommending API usage examples for different scenarios (Chap. 7), and one for further improving on existing snippets using information from question-answering services (Chap. 8).

Up until this point, we have mainly discussed the functional aspects of software development. To ensure, however, that reusing a component does not threaten the overall quality of the project under development, one must have certain indications that the component is of sufficient quality. We illustrate this need by means of a recommendation system for software components, which are assessed both from a functional and from a quality perspective (Chap. 9). The knowledge extraction layer in this case involves tools for extracting metrics from source code, documentation, version control systems, etc. Our focus is on static analysis metrics, for which we propose a reusability assessment approach that is constructed using as ground truth the popularity and the preference of software components, as measured by online services (Chap. 10).

2.4.4 An Integrated View of Software Reuse

As one may notice in the previous subsection, the contributions of the book can be seen as a series of different reuse options in each phase of software engineering. Although we do not claim to fully cover the reuse spectrum for software engineering, we argue that our solution can offer a useful set of solutions when it comes to building software products. In this subsection, we discuss this set of solutions, along with providing an integrated example on how we aspire our contributions to be used in the reuse context.

Assuming a team of engineers decides to build a new product, they would initially craft the requirements in cooperation with the relevant stakeholders. As an example, let's consider that the team wants to build a social network for academics. The reusable artifacts in this phase would most probably include functional requirements and UML diagrams. The requirements would include the actors of the system (e.g., academics), the main objects (e.g., publications, universities, etc.), and several action scenarios (e.g., academic logs in to the system, academic publishes paper, academic follows academic, etc.). The methodology presented in Part II of this book covers modeling this information (Chap. 3) and providing recommendations (in Chap. 4, several forgotten requirements can be recommended based on existing ones; e.g., given a scenario where an academic uploads a preprint, we could further suggest providing an option for editing the uploaded document).

Upon finalizing requirements elicitation and specification extraction, the development team should start developing the application. Source code development is of course a time-consuming process; they would initially have to design the required components, write code to develop the components, and integrate them with one another. Insights on source code component reuse are provided in Part III. In our social network example, the team would probably first have to develop some objects, such as a user account object and a publication object. Should they use our set of tools, they could find ready-to-use code for these objects. For instance, as the reader will see in Chaps. 5 and 6, the team could easily find a user account object by specifying its possible structure (i.e., variables, such as username or password, and/or methods, such as login, etc.). Issues related to source code, however, do not end here. The development of the project would require integrating the components, writing algorithms, connecting to external sources, debugging, etc. For instance, the engineers may need to find out how to connect a database to store the data of their social network (and also decide on which API to use for this task), or they may even have to determine whether the source code implementations of an algorithm could be improved. These are all queries that we aspire to be answered by the rest of Part III of this book (Chaps. 7 and 8).

Finally, before reusing (or after adopting and adapting) any components, the development team may also need to know if the components at hand are appropriate from a quality perspective. For instance, given a set of different components for a query about a user account object that are functionally equivalent, they may have to determine which one is the most reusable. This is a challenge for Part IV of this book, which proposes not only retrieving functionally useful components but also assessing their reusability using an estimation model (Chaps. 9 and 10). Using our solutions in this case, the team would not only save valuable time and effort, but they would also be more confident that the quality of the end product is acceptable.

References

1. Sommerville I (2010) Software engineering, 9th edn. Addison-Wesley, Harlow
2. IEEE (1993) IEEE standards collection: software engineering, IEEE standard 610.12-1990. Technical report, IEEE
3. Pressman R, Maxim B (2019) Software engineering: a practitioner's approach, 9th edn. McGraw-Hill Inc, New York
4. Pfleeger SL, Atlee JM (2009) Software engineering: theory and practice, 4th edn. Pearson, London
5. Dusink L, van Katwijk J (1995) Reuse dimensions. SIGSOFT Softw Eng Notes, 20(SI):137–149
6. Sametinger J (1997) Software engineering with reusable components. Springer-Verlag New York Inc, New York
7. Krueger CW (1992) Software reuse. ACM Comput Surv 24(2):131–183
8. Capilla R, Gallina B, Cetina C, Favaro J (2019) Opportunities for software reuse in an uncertain world: from past to emerging trends. J Softw: Evol Process 31(8):e2217
9. Arango GF (1988) Domain engineering for software reuse. PhD thesis, University of California, Irvine. AAI8827979
10. Prieto-Diaz R, Arango G (1991) Domain analysis and software systems modeling. IEEE Computer Society Press, Washington
11. Frakes W, Prieto-Diaz R, Fox C (1998) DARE: domain analysis and reuse environment. Ann Softw Eng 5:125–141
12. Czarnecki K, Østerbye K, Völter M (2002) Generative programming. In: Object-oriented technology ECOOP 2002 workshop reader, Berlin, Heidelberg. Springer, Berlin, Heidelberg, pp 15–29
13. Sillitti A, Vernazza T, Succi G (2002) Service oriented programming: a new paradigm of software reuse. In: Proceedings of the 7th international conference on software reuse: methods, techniques, and tools (ICSR-7), Berlin, Heidelberg. Springer, pp 269–280
14. Robillard MP, Maalej W, Walker RJ, Zimmermann T (2014) Recommendation Systems in Software Engineering. Springer Publishing Company, Incorporated, Berlin
15. Dagenais B, Ossher H, Bellamy RKE, Robillard MP, de Vries JP (2010) Moving into a new software project landscape. In: Proceedings of the 32nd ACM/IEEE international conference on software engineering - Volume 1 (ICSE'10), New York, NY, USA. ACM, pp 275–284
16. Sim SE, Gallardo-Valencia RE (2015) Finding source code on the web for remix and reuse. Springer Publishing Company, Incorporated, Berlin
17. Brandt J, Guo PJ, Lewenstein J, Dontcheva M, Klemmer SR (2009) Two studies of opportunistic programming: interleaving web foraging, learning, and writing code. In: Proceedings of the SIGCHI conference on human factors in computing systems (CHI'09), New York, NY, USA. ACM, pp 1589–1598
18. Li H, Xing Z, Peng X, Zhao W (2013) What help do developers seek, when and how? In: 2013 20th working conference on reverse engineering (WCRE), pp 142–151
19. Atwood J (2008) Introducing stackoverflow.com. https://blog.codinghorror.com/introducing-stackoverflow-com/. Accessed Nov 2017
20. Franck S (2008) None of us is as dumb as all of us. https://blog.codinghorror.com/stack-overflow-none-of-us-is-as-dumb-as-all-of-us/. Accessed Nov 2017

21. Garvin DA (1984) What does 'product quality' really mean? MIT Sloan Manag Rev 26(1)
22. Pirsig RM (1974) Zen and the art of motorcycle maintenance: an inquiry into values. Harper-Torch, New York
23. Suryn W (2014) Software quality engineering: a practitioner's approach. Wiley-IEEE Press, Hoboken
24. Pfleeger SL, Kitchenham B (1996) Software quality: the elusive target. IEEE Softw:12–21
25. Mccall JA, Richards PK, Walters GF (1977) Factors in software quality. Volume I. Concepts and definitions of software quality. Technical report ADA049014, General Electric Co, Sunnyvale, CA
26. Mccall JA, Richards PK, Walters GF (1977) Factors in software quality. Volume II. Metric data collection and validation. Technical report ADA049014, General Electric Co, Sunnyvale, CA
27. Mccall JA, Richards PK, Walters GF (1977) Factors in software quality. Volume III. Preliminary handbook on software quality for an acquisiton manager. Technical report ADA049014, General Electric Co, Sunnyvale, CA
28. ISO/IEC (1991) ISO/IEC 9126:1991. Technical report, ISO/IEC
29. ISO/IEC 25010:2011 (2011) https://www.iso.org/standard/35733.html. Accessed Nov 2017
30. Kan SH (2002) Metrics and models in software quality engineering, 2nd edn. Addison-Wesley Longman Publishing Co, Inc, Boston
31. Fenton N, Bieman J (2014) Software metrics: a rigorous and practical approach, 3rd edn. CRC Press Inc, Boca Raton
32. Bourque P, Fairley RE (eds) (2014) SWEBOK: guide to the software engineering body of knowledge, version 3.0 edition. IEEE Computer Society, Los Alamitos, CA
33. Thomas JM (1976) A complexity measure. In: Proceedings of the 2nd international conference on software engineering (ICSE'76), Los Alamitos, CA, USA. IEEE Computer Society Press, p 407
34. Halstead MH (1977) Elements of software science (operating and programming systems series). Elsevier Science Inc, New York
35. Chidamber SR, Kemerer CF (1994) A metrics suite for object oriented design. IEEE Trans Softw Eng 20(6):476–493
36. Henderson-Sellers B (1996) Object-oriented metrics: measures of complexity. Prentice-Hall Inc, Upper Saddle River
37. e Abreu FB, Carapuça R (1994) Candidate metrics for object-oriented software within a taxonomy framework. J Syst Softw 26(1):87–96
38. Lorenz M, Kidd J (1994) Object-oriented software metrics: a practical guide. Prentice-Hall Inc, Upper Saddle River
39. Srinivasan KP, Devi T (2014) A comprehensive review and analysis on object-oriented software metrics in software measurement. Int J Comput Sci Eng 6(7):247
40. Chhabra JK, Gupta V (2010) A survey of dynamic software metrics. J Comput Sci Technol 25(5):1016–1029
41. Yacoub SM, Ammar HH, Robinson T (1999) Dynamic metrics for object oriented designs. In: Proceedings of the 6th international symposium on software metrics (METRICS'99), Washington, DC, USA. IEEE Computer Society, p 50
42. Arisholm E, Briand LC, Foyen A (2004) Dynamic coupling measurement for object-oriented software. IEEE Trans Softw Eng 30(8):491–506
43. Schneidewind N (2009) Systems and software engineering with applications. Institute of Electrical and Electronics Engineers, New York
44. Albrecht AJ (1979) Measuring application development productivity. In: I. B. M. Press (ed) Proceedings of the IBM application development symposium, pp 83–92
45. Jones TC (1998) Estimating software costs. McGraw-Hill, Inc, Hightstown
46. Moser R, Pedrycz W, Succi G (2008) A comparative analysis of the efficiency of change metrics and static code attributes for defect prediction. In: Proceedings of the 30th international conference on software engineering (ICSE'08), New York, NY, USA. ACM, pp 181–190
47. Hassan AE (2009) Predicting faults using the complexity of code changes. In: Proceedings of the 31st international conference on software engineering (ICSE'09), Washington, DC, USA. IEEE Computer Society, pp 78–88

48. Spinellis D (2006) Code quality: the open source perspective. Effective software development series. Addison-Wesley Professional, Boston
49. BW Boehm (1981) Software engineering economics, 1st edn. Prentice Hall PTR, Upper Saddle River
50. Myers GJ, Sandler C, Badgett T (2011) The art of software testing, 3rd edn. Wiley Publishing, Hoboken
51. Beizer B (1990) Software testing techniques, 2nd edn. Van Nostrand Reinhold Co, New York
52. Copeland L (2003) A practitioner's guide to software test design. Artech House Inc, Norwood
53. Beck K (2002) Test driven development: by example. Addison-Wesley Longman Publishing Co, Inc, Boston
54. Beck K (2000) Extreme programming explained: embrace change. Addison-Wesley Longman Publishing Co, Inc, Boston
55. McConnell S (2004) Code complete, 2nd edn. Microsoft Press, Redmond
56. Witten IH, Frank E, Hall MA (2011) Data mining: practical machine learning tools and techniques, 3rd edn. Morgan Kaufmann Publishers Inc, San Francisco
57. Tan P-N, Steinbach M, Kumar V (2005) Introduction to data mining, 1st edn. Addison-Wesley Longman Publishing Co, Inc, Boston
58. Hand DJ, Smyth P, Mannila H (2001) Principles of data mining. MIT Press, Cambridge
59. Mitchell TM (1997) Machine learning, 1st edn. McGraw-Hill Inc, New York
60. Bishop CM (2006) Pattern recognition and machine learning. Springer-Verlag New York Inc, Secaucus
61. Manning CD, Raghavan P, Schütze H (2008) Introduction to information retrieval. Cambridge University Press, New York
62. Engelbrecht AP (2007) Computational intelligence: an introduction, 2nd edn. Wiley Publishing, Hoboken
63. Han J, Pei J, Kamber M (2011) Data mining: concepts and techniques. Morgan Kaufmann Publishers Inc, San Francisco
64. Delen D, Demirkan H (2013) Data, information and analytics as services. Decis Support Syst 55(1):359–363
65. Hastie T, Tibshirani R, Friedman J (2001) The elements of statistical learning. Springer series in statistics. Springer New York Inc, New York
66. Webb AR (2002) Statistical pattern recognition, 2 edn. Wiley, Hoboken
67. Urbanowicz RJ, Browne WN (2017) Introduction to learning classifier systems. Springerbriefs in intelligent systems. Springer New York Inc, New York
68. Xie T, Thummalapenta S, Lo D, Liu C (2009) Data mining for software engineering. Computer 42(8):55–62
69. Monperrus M (2013) Data-mining for software engineering. https://www.monperrus.net/martin/data-mining-software-engineering. Accessed Nov 2017
70. Halkidi M, Spinellis D, Tsatsaronis G, Vazirgiannis M (2011) Data mining in software engineering. Intell Data Anal 15(3):413–441
71. Kumar M, Ajmeri N, Ghaisas S (2010) Towards knowledge assisted agile requirements evolution. In: Proceedings of the 2nd international workshop on recommendation systems for software engineering (RSSE'10), New York, NY, USA. ACM, pp 16–20
72. Ghaisas S, Ajmeri N (2013) Knowledge-assisted ontology-based requirements evolution. In: Maalej W, Thurimella AK (eds) Managing requirements knowledge, pp 143–167. Springer, Berlin
73. Chen K, Zhang W, Zhao H, Mei H (2005) An approach to constructing feature models based on requirements clustering. In: Proceedings of the 13th IEEE international conference on requirements engineering (RE'05), Washington, DC, USA. IEEE Computer Society, pp 31–40
74. Alves V, Schwanninger C, Barbosa L, Rashid A, Sawyer P, Rayson P, Pohl C, Rummler A (2008) An exploratory study of information retrieval techniques in domain analysis. In: Proceedings of the 2008 12th international software product line conference (SPLC'08), Washington, DC, USA. IEEE Computer Society, pp 67–76

75. Felfernig A, Schubert M, Mandl M, Ricci F, Maalej W (2010) Recommendation and decision technologies for requirements engineering. In: Proceedings of the 2nd international workshop on recommendation systems for software engineering (RSSE'10), New York, NY, USA. ACM, pp 11–15

76. Maalej W, Thurimella AK (2009) Towards a research agenda for recommendation systems in requirements engineering. In: Proceedings of the 2009 2nd international workshop on managing requirements knowledge (MARK'09), Washington, DC, USA. IEEE Computer Society, pp 32–39

77. Binkley D (2007) Source code analysis: a road map. In: 2007 Future of software engineering (FOSE'07), Washington, DC, USA. IEEE Computer Society, pp 104–119

78. Janjic W, Hummel O, Schumacher M, Atkinson C (2013) An unabridged source code dataset for research in software reuse. In: Proceedings of the 10th working conference on mining software repositories (MSR'13), Piscataway, NJ, USA. IEEE Press, pp 339–342

79. Bajracharya S, Ngo T, Linstead E, Dou Y, Rigor P, Baldi P, Lopes C (2006) Sourcerer: a search engine for open source code supporting structure-based search. In: Companion to the 21st ACM SIGPLAN symposium on object-oriented programming systems, languages, and applications (OOPSLA'06), New York, NY, USA. ACM, pp 681–682

80. Linstead E, Bajracharya S, Ngo T, Rigor P, Lopes C, Baldi P (2009) Sourcerer: mining and searching internet-scale software repositories. Data Min Knowl Discov 18(2):300–336

81. Holmes R, Murphy GC (2005) Using structural context to recommend source code examples. In: Proceedings of the 27th international conference on software engineering (ICSE'05), New York, NY, USA. ACM, pp 117–125

82. Mandelin D, Lin X, Bodík R, Kimelman D (2005) Jungloid mining: helping to navigate the API jungle. SIGPLAN Not 40(6):48–61

83. Thummalapenta S, Xie T (2007) PARSEWeb: a programmer assistant for reusing open source code on the web. In: Proceedings of the 22nd IEEE/ACM international conference on automated software engineering (ASE'07), New York, NY, USA. ACM, pp 204–213

84. Hummel O, Janjic W, Atkinson C (2008) Code conjurer: pulling reusable software out of thin air. IEEE Softw 25(5):45–52

85. Lazzarini Lemos OA, Bajracharya SK, Ossher J (2007) CodeGenie: a tool for test-driven source code search. In: Companion to the 22nd ACM SIGPLAN conference on object-oriented programming systems and applications companion (OOPSLA'07), New York, NY, USA. ACM, pp 917–918

86. Reiss SP (2009) Semantics-based code search. In: Proceedings of the 31st international conference on software engineering (ICSE'09), Washington, DC, USA. IEEE Computer Society, pp 243–253

87. Hummel O, Atkinson C (2004) Extreme harvesting: test driven discovery and reuse of software components. In: Proceedings of the 2004 IEEE international conference on information reuse and integration (IRI 2004), pp 66–72

88. Nurolahzade M, Walker RJ, Maurer F (2013) An assessment of test-driven reuse: promises and pitfalls. In: Favaro J, Morisio M (eds) Safe and secure software reuse. Lecture notes in computer science, vol 7925. Springer, Berlin, pp 65–80

89. Xie T, Pei J (2006) MAPO: mining API usages from open source repositories. In: Proceedings of the 2006 international workshop on mining software repositories (MSR'06), New York, NY, USA. ACM, pp 54–57

90. Sahavechaphan N, Claypool K (2006) XSnippet: mining for sample code. SIGPLAN Not 41(10):413–430

91. Zagalsky A, Barzilay O, Yehudai A (2012) Example overflow: using social media for code recommendation. In: Proceedings of the 3rd international workshop on recommendation systems for software engineering (RSSE'12), Piscataway, NJ, USA. IEEE Press, pp 38–42

92. Wightman D, Ye Z, Brandt J, Vertegaal R (2012) SnipMatch: using source code context to enhance snippet retrieval and parameterization. In: Proceedings of the 25th annual ACM symposium on user interface software and technology (UIST'12), New York, NY, USA. ACM, pp 219–228

93. Wei Y, Chandrasekaran N, Gulwani S, Hamadi Y (2015) Building bing developer assistant. Technical report MSR-TR-2015-36, Microsoft Research
94. Kagdi H, Collard ML, Maletic JI (2007) A survey and taxonomy of approaches for mining software repositories in the context of software evolution. J Softw Maint Evol 19(2):77–131
95. Alves TL, Ypma C, Visser J (2010) Deriving metric thresholds from benchmark data. In: Proceedings of the IEEE international conference on software maintenance (ICSM). IEEE, pp 1–10
96. Ferreira KAM, Bigonha MAS, Bigonha RS, Mendes LFO, Almeida HC (2012) Identifying thresholds for object-oriented software metrics. J Syst Softw 85(2):244–257
97. Zhong S, Khoshgoftaar TM, Seliya N (2004) Unsupervised learning for expert-based software quality estimation. In: Proceedings of the 8th IEEE international conference on high assurance systems engineering (HASE'04), pp 149–155
98. Hovemeyer D, Spacco J, Pugh W (2005) Evaluating and tuning a static analysis to find null pointer bugs. SIGSOFT Softw Eng Notes 31(1):13–19
99. Ayewah N, Hovemeyer D, Morgenthaler JD, Penix J, Pugh W (2008) Using static analysis to find bugs. IEEE Softw 25(5):22–29
100. Le Goues C, Weimer W (2012) Measuring code quality to improve specification mining. IEEE Trans Softw Eng 38(1):175–190
101. Washizaki H, Namiki R, Fukuoka T, Harada Y, Watanabe H (2007) A framework for measuring and evaluating program source code quality. In: Proceedings of the 8th international conference on product-focused software process improvement (PROFES). Springer, pp 284–299
102. Cai T, Lyu MR, Wong K-F, Wong M (2001) ComPARE: a generic quality assessment environment for component-based software systems. In: Proceedings of the 2001 international symposium on information systems and engineering (ISE'2001)

Part II
Requirements Mining

Chapter 3
Modeling Software Requirements

3.1 Overview

Typically, during the early stages of the software development life cycle, developers, and customers discuss and agree on the functionality of the system to be developed. The functionality is recorded as a set of functional requirements, which form the basis for the corresponding work implementation plan, cost estimations, and follow-up directives [1]. Functional requirements can be expressed in various ways, including UML diagrams, storyboards, and, most often, natural language [2].

Deriving formal specifications from functional requirements is one of the most important steps of the software development process. The source code of a project usually depends on an initial model (e.g., described by a class diagram) that has to be designed very thoroughly in order to be functionally complete. Designing such a model from scratch is not a trivial task. While requirements expressed in natural language have the advantage of being intelligible to both clients and developers, they can also be ambiguous, incomplete, and inconsistent. Several formal languages have been proposed to eliminate some of these problems; however, customers rarely possess the technical expertise for constructing and understanding highly formalized requirements. Thus, investing effort into automating the process of translating requirements to specifications can be highly cost-effective, as it removes the need for customers to understand complex design models.

Automating the requirements translation process allows detecting ambiguities and errors at an early stage of the development process (e.g., via logical inference and verification tools), thus avoiding the costs of finding and fixing problems at a later, more expensive stage [3]. In the field of requirements modeling, there are several techniques for mapping requirements to specifications; however, they mostly depend on domain-specific heuristics and/or controlled languages. In this chapter, we propose a methodology that allows developers to design their envisioned system through software requirements in multimodal formats. In specific, our system effectively models the static and dynamic views of a software project, and employs NLP

© Springer Nature Switzerland AG 2020
T. Diamantopoulos and A. L. Symeonidis, *Mining Software Engineering Data for Software Reuse*, Advanced Information and Knowledge Processing, https://doi.org/10.1007/978-3-030-30106-4_3

Fig. 3.1 Example scenario
using the language of project
ReDSeeDS [4]

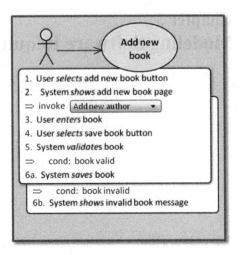

techniques and semantics to transform user requirements to system specifications.
The derived representations are instantiated in software ontologies using tools specif-
ically designed and developed for generating models from functional requirements,
UML diagrams, and storyboards (i.e., graphical scenarios).

Note that the methodology proposed in this chapter is applicable to various software engineering processes and scenarios. We provide an indicative application of
our methodology on RESTful web service development and demonstrate how the
construction of service prototypes can be greatly facilitated by providing fully traceable and effective specifications for a web service.

3.2 State of the Art on Requirements Elicitation and Specification Extraction

Translating user requirements to specifications involves designing meaningful representations and instantiating them using input from user requirements. Concerning
the type of the input, most approaches propose specialized languages that can be easily translated to models. One of the most representative examples is the *ReDSeeDS*
project [4, 5] that uses a constrained language named *Requirements Specification
Language (RSL)* [4] to extract specifications from use cases and domain models.
Figure 3.1 depicts an RSL example for adding a new book to a library management
system. Note that the RSL is clearly structured (i.e., the support for natural language
is very limited); however, it also supports invoking external scenarios, as well as
branching (e.g., as in the figure where the response depends on the validity of the
book).

Cucumber [6] and *JBehave* [7] are also popular frameworks based on the
Behavior-Driven Development (BDD) paradigm. They allow defining scenarios using

```
public class TraderSteps {
    private TradingService service;    // Injected
    private Stock stock; // Created

    @Given("a stock and a threshold of $threshold")
    public void aStock(double threshold) {
        stock = service.newStock("STK", threshold);
    }
    @When("the stock is traded at price $price")
    public void theStockIsTraded(double price) {
        stock.tradeAt(price);
    }
    @Then("the alert status is $status")
    public void theAlertStatusIs(String status) {
        assertThat(stock.getStatus().name(), equalTo(status));
    }
}
```

```
Scenario: A trader is alerted of status

Given a stock and a threshold of 15.0
When stock is traded at 5.0
Then the alert status should be OFF
When stock is traded at 16.0
Then the alert status should be ON
```

Fig. 3.2 Example scenario and code skeleton using JBehave [7]

a *given-when-then* approach and aspire to construct testable behavioral models. The *given* part of the scenario includes any initializations and/or preconditions, while the *when-then* structure typically includes responses to different inputs to the system. Figure 3.2 depicts a stock market trading service scenario built using JBehave. The scenario, which is written using the Cucumber *given-when-then* approach, is shown in the left part of the figure. The right part illustrates how the tool extracts specifications as objects and even supports producing skeleton code. Finally, many approaches use specialized languages like *Tropos*, a requirements-driven methodology designed by Mylopoulos et al. [8], which is based on the notions of actors and goals of the i* modeling framework [9].

Although the aforementioned approaches can be quite useful for building precise models, their applicability is limited, since using them requires training in a new language and/or framework, which is sometimes frustrating for the developers. To this end, current literature proposes employing semantics and NLP techniques on natural (or semi-structured) language functional requirements and UML diagrams to extract specifications. One of the early rule-based methods for extracting data types, variables, and operators from requirements was introduced by Abbott [10], according to which nouns are identified as objects and verbs as operations between them. Subsequently, Abbott's approach was extended to object-oriented development by Booch [11].

Saeki et al. [12] were among the first to construct a system for extracting object-oriented models from informal requirements. Their system uses NLP methods to extract nouns and verbs from functional requirements and determines whether they are relevant to the model by means of human intervention. The work by Mich [13] also involves a semantics module using a knowledge base in order to further assess the retrieved syntactic terms. Alternatively, Harmain and Gaizauskas [14] developed *CM-Builder*, a natural language upper CASE tool in order to identify object classes, attributes, and relationships in the analysis stage using functional requirements. CM-Builder initially employs a parser to extract noun phrases and verb phrases and generates a list of candidate classes and attributes. After that, the tool also provides candidate relationships and requires the user to examine them. An example for a

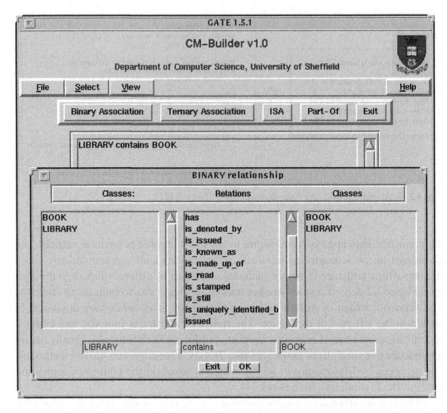

Fig. 3.3 Example relationship between classes using CM-Builder [14]

candidate relationship between a library class and a book class is shown in Fig. 3.3. In this example, the user has selected the relationship library-contains-book.

Finally, there are several commercial solutions for requirements authoring and validation. For example, the Reuse Company[1] offers solutions for modeling requirements using structured patterns and further assists in maintaining a common terminology throughout the domain of the project. A similar solution is offered by Innoslate,[2] which further provides version control for requirements. These tools are usually domain-specific, enabling reuse within a project, and allow extracting entities for validation purposes.

Concerning model representations, most of the aforementioned systems use either known literature standards including UML models or ontologies, which have been used extensively in *Requirements Engineering (RE)* for storing and validating information extracted from requirements [15, 16]. Ontologies are also useful for storing domain-specific knowledge and are known to integrate well with software modeling

---

[1]https://www.reusecompany.com/.

[2]https://www.innoslate.com/.

languages [17].[3] They provide a structured means of organizing information, are particularly useful for storing linked data, and allow reasoning over implied relationships between data items.

Though effective, these methodologies are usually restricted to formalized languages or depend on heuristics and domain-specific information. To this end, we propose to employ semantic role labeling techniques in order to learn to map functional requirements concepts to formal specifications. Note that although extracting semantic relations between entities is one of the most well-known problems of NLP,[4] applying these methods to the analysis of user requirements is a rather new idea.

Furthermore, with respect to current methodologies, we also aspire to provide a unified model for storing requirements from multimodal formats, taking into account both the static and dynamic views of software projects. Our model may be instantiated from functional requirements written in natural language, UML use case and activity diagrams, and graphical storyboards, for which we have constructed appropriate tools. The extracted information is stored into ontologies, for analysis and reuse purposes.

3.3 From Requirements to Specifications

3.3.1 System Overview

Our approach comprises the *Reqs2Specs* module, and its conceptual architecture is shown in Fig. 3.4. The Reqs2Specs module includes a set of tools for developers to enter the multimodal representations (functional requirements, storyboards) of the envisaged software application as well as a methodology for transforming these representations to specifications, i.e., models of the envisioned software project.

We have developed tools that allow developers to insert requirements in the form of semi-structured text, UML use case and activity diagrams, and storyboards, i.e., diagrams depicting the flow of actions in the system. Using our approach, these artifacts are syntactically and semantically annotated and stored in two ontologies, the static ontology and the dynamic ontology, which practically store the static elements and the dynamic elements of the system.

Having finalized the requirements elicitation process, the Reqs2Specs module parses the ontologies into an aggregated ontology and extracts detailed system

---

[3] See [18] for a systematic review on the use of ontologies in RE.

[4] There are several approaches that can be distinguished into feature-based and kernel-based methods. Given a dataset with annotated relations, feature-based methods involve extracting syntactic and semantic features from text, predefined accordingly, and providing the feature set to a classifier which is trained in order to identify relations [19–21]. Kernel-based methods, on the other hand, do not require manual feature extraction as they map text entities and relations in a higher dimensional representation, such as bag-of-words kernels [22] or tree kernels [23–25], and classify relations according to their new representation. Both types of methods have been found to be effective under different scenarios [26].

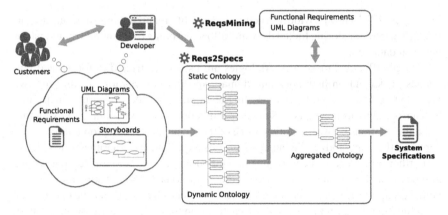

Fig. 3.4 Overview of the conceptual architecture of our system

specifications, which are suitable for use either directly by developers or even as input in an automated source code generation engine (as performed in S-CASE [27]). The ReqsMining module is used to provide recommendations about reusable information present in functional requirements and UML diagrams. The methodologies employed in this module are analyzed in detail in the next chapter. Note also that our modeling methodology is fully traceable; any changes to the ontologies (e.g., as part of a mining process) can be propagated back to the functional requirements, UML diagrams, and storyboards. As a result, the stakeholders are always provided with a clear requirements view of the system, and the specifications are well defined. The following subsections outline the developed ontologies and the process of instantiating them from functional requirements, UML diagrams, and storyboards.

3.3.2 Extracting Artifacts from Software Requirements

This section discusses the design of the two ontologies that store information capturing the static view (functional requirements, UML use case diagrams) and the dynamic view (storyboards, UML activity diagrams) of software projects. These ontologies provide a structured means of organizing information, underpin methods for retrieving stored data via queries, and allow reasoning over implied relationships between data items.

The Resource Description Framework (RDF)[5] was selected as the formal framework for representing requirements-related information. The RDF data model has three object types: *resources*, *properties*, and *statements*. A resource is any "thing" that can be described by the language, while properties are binary relations. An RDF statement is a triple consisting of a resource (the subject), a property, and either a

---

[5]http://www.w3.org/TR/1999/REC-rdf-syntax-19990222.

resource or a string as object of the property. Since the RDF data model defines no syntax for the language, RDF models can be expressed in different formats; the two most prevalent are XML and Turtle.[6] RDF models are often accompanied by an RDF schema (RDFS)[7] that defines a domain-specific vocabulary of resources and properties. Although RDF is an adequate basis for representing and storing information, reasoning over the information and defining an explicit semantics for it is hard. RDFS provides some reasoning capabilities but is intentionally inexpressive, and fails to support the logical notion of negation. For this purpose, the Web Ontology Language (OWL)[8] was designed as a more expressive formalism that allows classes to be defined axiomatically and supports consistency reasoning over classes. OWL is built on top of RDF but includes a richer syntax with features such as cardinality or inverse relations between properties. In the context of our work, we employ OWL for representing information, since it is a widely used standard within the research and industry communities.[9]

3.3.2.1 Static View of Software Projects

An Ontology for the Static View of Software Projects Concerning the static aspects of requirements elicitation, i.e., functional requirements and use case diagrams, the design of the ontology revolves around acting units (e.g., user, administrator, guest) performing some action(s) on some object(s). The ontology was designed to support information extracted from Subject-Verb-Object (SVO) sentences. The class hierarchy of the ontology is shown in Fig. 3.5.

Anything entered in the ontology is a Concept. Instances of Concept are further classified into Project, Requirement, ThingType, and OperationType. The Project and Requirement classes are used to store the requirements of the system, so that any instance can be traced back to the originating requirement. ThingType and OperationType are the main types of objects found in any requirement. ThingType refers to acting units and units acted upon, while OperationType involves the types of actions performed by the acting units. Each ThingType instance can be an Actor, an Object, or a Property. Actor refers to the actors of the project, including users, the system itself, or any external systems. Instances of type Object include any object or resource of the system that receives some action, while Property instances include all modifiers of objects or actions. OperationType includes all possible operations, including Ownership that expresses possession (e.g., "each user has an account"), Emergence that implies passive transformation (e.g., "the posts are

---

[6]http://www.w3.org/TR/turtle/.

[7]http://www.w3.org/TR/1999/PR-rdf-schema-19990303/.

[8]http://www.w3.org/TR/2004/REC-owl-guide-20040210/.

[9]Although we don't rely on OWL inference capabilities explicitly, they are useful for expressing integrity conditions over the ontology, such as ensuring that certain properties have inverse properties (e.g., owns/owned_by).

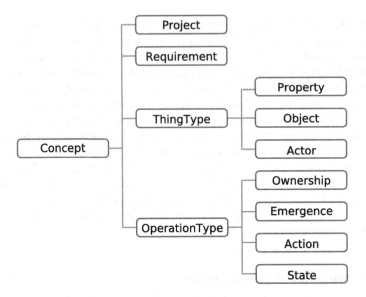

Fig. 3.5 Static ontology of software projects

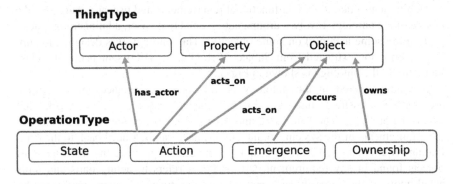

Fig. 3.6 Properties of the static ontology

sorted"), State that describes the status of an Actor (e.g., "the user is logged in"), and Action describes an operation performed by an Actor on some Object (e.g., "the user creates a profile").

The possible interactions between the different concepts of the ontology, i.e., the relations between the ontology (sub)classes, are defined using *properties*. The properties of the static ontology are shown in Fig. 3.6. Note that each property also has its inverse one (e.g., has_actor has the inverse is_actor_of). Figure 3.6 includes only one of the inverse properties and excludes properties involving Project and Requirement for simplicity.

An instance of Project can have many instances of Requirement, while each Requirement connects to several ThingType and OperationType

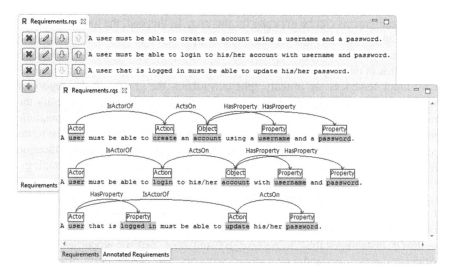

Fig. 3.7 Screenshot of the requirements editor

instances. The remaining properties define relations between these instances. `OperationType` instances connect with instances of `Actor` via the `has_actor` property. Operations also connect to objects if they are transitive. Thus, each `Action` is connected to instances of type `Object` or `Property` via the `acts_on` property, while `Emergence occurs` on an `Object` and `Ownership` is connected with objects via the `owns` and `owned_by` properties. Finally, the non-transitive `State` operation does not connect to any `Object` or `Property`.

Extracting Artifacts from Functional Requirements The ontology described in the previous paragraph is instantiated using functional requirements. The requirements have to be *annotated*, and the annotations are used to map to the OWL classes of the ontology. Thus, we designed an annotation scheme that includes four types of entities and three types of relations among them. In specific, any entity can be an *Actor*, an *Action*, an *Object*, or a *Property*. The defined relations include *IsActorOf* declared from *Actor* to *Action*, *ActsOn* defined from *Action* to *Object* or from *Action* to *Property*, and *HasProperty* defined from *Actor* to *Property* or from *Object* to *Property* or from *Property* to *Property*. Thus, using this annotation scheme, we have developed a tool for adding, modifying, and annotating functional requirements. The *Requirements Editor* is built as an Eclipse plugin using the SWT and is shown in Fig. 3.7.

On the first page of the tool (shown in the top left of Fig. 3.7), the user can add, delete, or modify functional requirements. The second page of the editor (shown in the bottom right of Fig. 3.7) refers to the annotations of the requirements. The supported functionality includes adding and deleting entity and relationship annotations.

Given that the structure of functional requirements usually follows the SVO motif (as in Fig. 3.7), annotating them is intuitive. However, the procedure of annotating several requirements can be tiresome for the user. To this end, the tool can be connected to an external NLP parser in order to support semi-automatically annotating requirements. We have constructed such an example parser (described in Sect. 3.3.3) that operates in two stages. The first stage is the *syntactic analysis* for each requirement. Input sentences are initially split into tokens, the grammatical category of these tokens is identified, and their base types are extracted, to finally identify the grammatical relations between the words. The second stage involves the *semantic analysis* of the parsed sentences. This stage extracts semantic features for the terms (e.g., part of speech, relation to parent lemma, etc.) and employs a classification algorithm to classify each term to the relevant concept or operation of the ontology.

Upon annotation, the user can select the option to export the annotations to the static ontology. The mapping from the annotation scheme to the concepts of the static ontology is straightforward. *Actor*, *Action*, *Object*, and *Property* annotations correspond to the relevant OWL classes. Concerning relations, *IsActorOf* instantiates the is_actor_of and has_actor properties, *ActsOn* instantiates acts_on and receives_action, and *HasProperty* maps to the has_property and is_property_of properties. The rest of the ontology properties (e.g., project_has, consists_of) are instantiated using the information of each functional requirement and of the name of the software project.

3.3.2.2 Dynamic View of Software Projects

An Ontology for the Dynamic View of Software Projects In this paragraph, we discuss the developed ontology that captures the dynamic view of a system. The main elements of dynamic representations are flows of actions among system objects. Using OWL, actions can be represented as classes and flows can be described using properties. The class hierarchy of the ontology is shown in Fig. 3.8.

Anything entered in the ontology is a Concept. Instances of class Concept are further divided into Project, ActivityDiagram, AnyActivity, Actor, Action, Object, Condition, Transition, and Property. The class Project refers to the project analyzed, while ActivityDiagram stores each diagram of the system, including not only activity diagrams, but also storyboards and generally any diagrams with dynamic flows of actions. As in the static ontology, Project and ActivityDiagram can be used to ensure that the concepts extracted from the ontology can be traced in the original diagram representations, allowing to reconstruct them.

Activities are the main building blocks of dynamic system representations. The activities of a diagram instantiate the OWL class AnyActivity. This class is further distinguished into the subclasses InitialActivity, FinalActivity, and Activity. InitialActivity refers to the initial state of the diagram, and FinalActivity refers to the final state of the diagram, while Activity holds any intermediate activities of the system. Any action of the system may also require one or more input properties, stored in class Property. For instance,

Fig. 3.8 Dynamic ontology of software projects

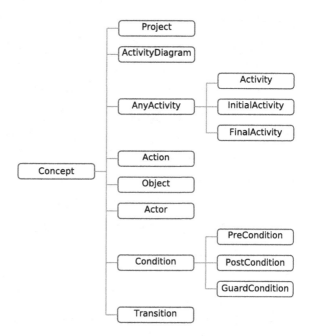

performing a "Create account" may require a "username" and a "password". In this case, "username" and "password" are instances of class `Property`.

The flow of activities in storyboards or activity diagrams is described using transitions. The OWL class `Transition` describes the flow from one instance of `Activity` to the next instance of `Activity` as derived by the corresponding diagram. Each `Transition` may also have a `Condition`. The class `Condition` has three subclasses: `PreCondition`, `PostCondition`, and `GuardCondition`. The first two refer to conditions that have to be met before (`PreCondition`) or after (`PostCondition`) the execution of the activity flow of the diagram, while `GuardCondition` is a condition that "guards" the execution of an activity of the system along with the corresponding answer. For example, "Create account" may be guarded by the condition "is the username unique? Yes", while the opposite `GuardCondition` "is the username unique? No" shall not allow executing the "Create account" activity.

The properties of the ontology define the possible interactions between the different classes, involving interactions at inter-diagram level and relations between elements of a diagram. The properties of the dynamic ontology are illustrated in Fig. 3.9, including only one of the inverse properties and excluding the `Project` and `ActivityDiagram` properties for simplicity. Each project can have one or more diagrams, and each diagram has to belong to a project. Additionally, each diagram may have a `PreCondition` and/or a `PostCondition`. An instance of `ActivityDiagram` has elements of the five classes `Actor`, `AnyActivity`, `Transition`, `Property`, and `Condition`.

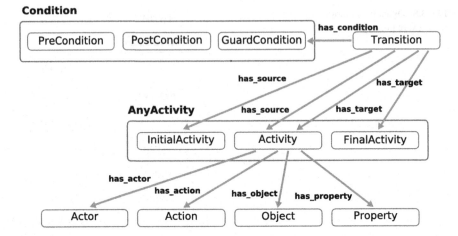

Fig. 3.9 Properties of the dynamic ontology

The different relations of ontology classes are actually forming the main flow as derived from diagram elements. Thus, Activity instances are connected with each other via instances of type Transition. Any Transition has a source and a target Activity (properties has_source and has_target, respectively), and it may also have a GuardCondition (property has_condition). Finally, any Activity is related to instances of type Property, while any GuardCondition has an opposite one, connected to each other via the bidirectional property is_opposite_of. Finally, any Activity is connected to an Actor, an Action, and an Object via the corresponding properties has_actor, activity_has_action, and activity_has_object.

Extracting Dynamic Flow from Storyboards Storyboards are dynamic system scenarios that describe flows of actions in software systems. A storyboard diagram is structured as a flow from a *start node* to an *end node*. Between the start and the end nodes, there are *actions* with their *properties* and *conditions*. All nodes are connected with edges/paths. We have designed and implemented *Storyboard Creator*, a tool for creating and editing storyboard diagrams, as an Eclipse plugin using the Graphical Modeling Framework (GMF). A screenshot of the tool including a storyboard is shown in Fig. 3.10.

The Storyboard Creator includes a canvas for drawing storyboards, a palette that can be used to create nodes and edges, and an outline view that can be used to scroll when the diagram does not fit into the canvas. It also supports validating diagrams using several rules, e.g., to ensure that each diagram has at least one action node, each property connects to exactly one action, etc.

The storyboard of Fig. 3.10 includes an action "Login to account" which also has two properties that define the elements of "account", a "username", and a "password". Conditions have two possible paths. For instance, condition "Credentials are

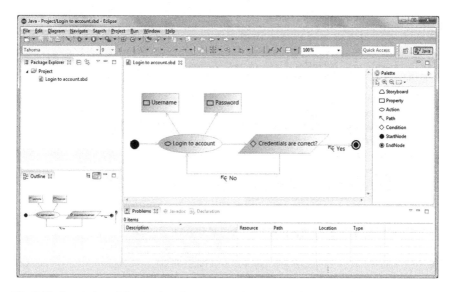

Fig. 3.10 Screenshot of the storyboard creator

Table 3.1 Example instantiated OWL classes for the storyboard of Fig. 3.10

OWL class	OWL instances
Activity	Login_to_account
Property	Username, Password
Transition	FROM__StartNode__TO__Login_to_account, FROM__Login_to_account__TO__EndNode, FROM__Login_to_account__TO__Login_to_account
GuardCondition	Credentials_are_correct__PATH__Yes, Credentials_are_correct__PATH__No
Action	login
Object	account

correct?" has the "Yes" path that leads to the end node and the "No" path that leads back to the login action requiring new credentials.

Mapping storyboard diagrams to the dynamic ontology is straightforward. Storyboard actions are mapped to the OWL class Activity, and they are further split into instances of Action and Object. Properties become instances of the Property class, and they are connected to the respective Activity instances via the has_property relation. The paths and the condition paths of storyboards become instances of Transition, while the storyboard conditions split into two opposite GuardConditions. An example instantiation for the storyboard of Fig. 3.10 is shown in Table 3.1.

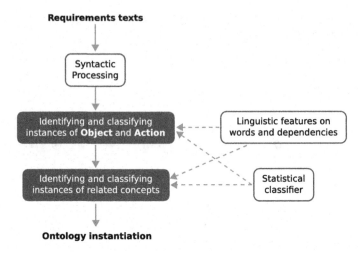

Fig. 3.11 Pipeline architecture for parsing software requirements

3.3.3 A Parser for Automated Requirements Annotation

Based on the modeling effort described in Sect. 3.3.2, we have developed a parser that
learns to automatically map software requirements written in natural language texts
to concepts and relations defined in the ontology. We identify instances of the concepts `Actor`, `Action`, `Object`, and `Property`, and the relations among them
(`is_actor_of`, `acts_on`, and `has_property`). These concepts are selected
since they align well with key syntactic elements of the sentences, and are intuitive
enough to make it possible for non-linguists to annotate them in texts. Our parser,
however, could also identify more fine-grained concepts, given more detailed annotations. In practice, the parsing task involves several steps: first, concepts need to
be identified and then mapped to the correct class and, second, relations between
concepts need to be identified and labeled accordingly. We employ a feature-based
relation extraction method, since we are focusing on a specific problem domain and
type of input, i.e., textual requirements of software projects. This allows us to implement the parsing pipeline shown in Fig. 3.11. In the following paragraphs, we provide
a summary of constructing, training, and evaluating our parser, while the interested
reader is further referred to [28] and [29] for the details of our methodology.

Our parser comprises two subsystems, one for syntactic and one for semantic
analysis. Syntactic analysis allows us to identify the grammatical category of each
word in a requirement sentence, while semantic analysis is used to map words and
constituents in a sentence to instances of concepts and relations from the ontology.

The syntactic analysis stage of our pipeline architecture performs the following
steps: tokenization, part-of-speech tagging, lemmatization, and dependency parsing.
Given an input sentence, this means that the pipeline separates the sentence into
word tokens, identifies the grammatical category of each word (e.g., "user" → noun,

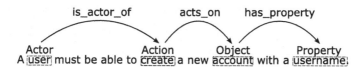

Fig. 3.12 Example annotated instance using the hierarchical annotation scheme

"create" → verb), and determines their uninflected base forms (e.g., "users" → "user"). Finally, the pipeline identifies the grammatical relations that hold between two words (e.g., ⟨"user", "must"⟩ → subject-of, ⟨"create", "account"⟩ → object-of). We construct the pipeline using the Mate Tools [30, 31],[10] which is a system that includes parsing components and pre-trained models for syntactic analysis.

The semantic analysis subsystem also adopts a pipeline architecture to receive the output of syntactic preprocessing and extract instances of ontology concepts and relations. The analysis is carried out in four steps: (1) identifying instances of Action and Object; (2) allocating these to the correct concept (either Action or Object); (3) identifying instances of related concepts (i.e., Actor and Property), and (4) determining their relationships to concept instances identified in step (1). These four steps correspond to the *predicate identification, predicate disambiguation, argument identification*, and *argument classification* phases of Mate Tools [30]. Our method is based on the semantic role labeler from Mate Tools and uses the built-in re-ranker to find the best joint output of steps (3) and (4). We extend Mate Tools with respect to continuous features and arbitrary label types. Each step in our pipeline is implemented as a logistic regression model using the LIBLINEAR toolkit [32] that employs linguistic properties as features, for which appropriate feature weights are learned based on annotated training data (see [28] and [29] for more information on building this example parser).

Training the parser requires a dataset of annotated functional requirements. Our annotation procedure is quite simple and involves marking instances of Actor, Object, Action, and Property that are explicitly expressed in a given requirement. Consider the example of Fig. 3.12. In this sentence, "user" is annotated as Actor, "create" as Action, "account" as Object, and "username" as Property.[11]

Since annotating can be a hard task, especially for inexperienced users, we created an annotation tool, named *Requirements Annotation Tool*, to ease the process. The Requirements Annotation Tool is a web platform that allows users to create an account, import one or more of their projects, and annotate them. Practically, the tool is similar to the Requirements Editor presented in Sect. 3.3.2; however, the main scope of the Requirements Annotation Tool is annotation for parser training, so it provides an intuitive drag n' drop environment. The two tools are compatible as they allow importing/exporting annotations using the same file formats.

---

[10]http://code.google.com/p/mate-tools/.

[11] More specific or implicit annotations could also be added using the hierarchical annotation scheme defined in [29].

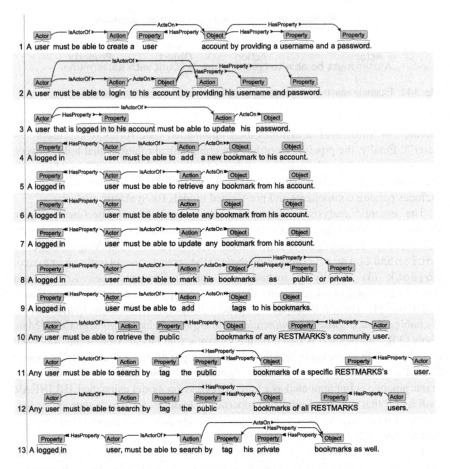

Fig. 3.13 Annotated requirements of project Restmarks

As an example of using our annotation tool, we use the requirements of project *Restmarks*. Restmarks is a demo project and can be seen as a social service for bookmarks. The users of the service can store and retrieve their bookmarks, share them with the community and search for bookmarks by using tags. A screenshot of the tool depicting the annotations for Restmarks is shown in Fig. 3.13.

The annotations are comprehensive; even users with no experience should be able to identify and label the appropriate entities and relations. The tool exports annotations in different formats, including OWL and TTL. Table 3.2 illustrates the Restmarks entities discovered via the tool.

The tool also instantiates the properties of the ontology. For example, the properties for requirement FR4 of Restmarks are shown in Table 3.3.

Given the annotation tool, we were able to construct a dataset of annotated requirements that were used to train our parser. In specific, we collected requirements

Table 3.2 Ontology instances for the entities of Restmarks

OWL class	OWL individual(s)
Project	Restmarks
Requirement	FR1, FR2, FR3, FR4, FR5, FR6, FR7, FR8, FR9, FR10, FR11, FR12, FR13
Actor	user, users
Action	create, retrieve, mark, search, delete, add, providing, login, update
Object	account, bookmark, bookmarks, password_1, tags
Property	RESTMARKS, password, tag, public, private, logged_in_to_his_account, logged_in, username, user_1

Table 3.3 Properties for the ontology instances of the FR4 of Restmarks

OWL individual	OWL property	OWL individual
user	is_actor_of	add
add	has_actor	user
add	acts_on	bookmark
bookmark	receives_action	add

of various domains; the dataset comprises requirements of documents written by students as part of university courses[12] as well as industrial prototypes, including RESTful prototype applications from project S-CASE.[13] All in all, we collected 325 requirements sentences, which, upon tokenization, resulted in 4057 tokens. After that, the sentences were annotated by two different annotators, and they were carefully examined one more time in order to produce a well-annotated dataset (see [29] for more details about the annotation process). The final dataset included 2051 class instances (435 actions, 305 actors, 613 objects, and 698 properties) and 1576 relations among them (355 has_actor relations, 531 acts_on relations, and 690 has_property relations).

As evaluation metrics for our parser, we apply labeled precision and recall. We define *labeled precision* as the fraction of predicted labels of concept and relation instances that are correct, and *labeled recall* as the fraction of annotated labels that are correctly predicted by the parser. To train and test the statistical models underlying the semantic analysis components of our pipeline, we perform evaluation in a fivefold cross-validation setting. That is, given the 325 sentences from the annotated dataset, we randomly create fivefold of equal size (65 sentences) and use each fold

---

[12]The majority of requirements collected in this way were provided by a software development course organized jointly by several European universities, which is available at http://www.fer.unizg.hr/rasip/dsd.

[13]https://s-case.github.io.

once for testing while training on the remaining other folds. Using this setting, our model achieves a precision and recall of 77.9% and 74.5%, respectively. As these metrics illustrate, our semantic parsing module is quite effective in terms of software engineering automation. In particular, the recall value indicates that our module correctly identifies roughly three out of four of all annotated instances and relations. Additionally, the precision of our module indicates that approximately four out of five annotations are correct. Thus, our parser significantly reduces the effort (and time) required to identify the concepts that are present in functional requirements and annotate them accordingly.

3.3.4   From Software Artifacts to Specifications

The ontologies developed in Sect. 3.3.2 constitute on their own a model that is suitable for storing requirements, as well as for actions such as requirements validation and mining (see following chapter). Their aggregation can lead to a final model that would describe the system under development both from a static and from a dynamic perspective. This model can be subsequently used to design the final system or even automatically develop it (as performed in [27]). Without loss of generality, we build an aggregated model for RESTful web services and construct a YAML representation that functions as the final model of the service under development.

3.3.4.1   Aggregated Ontology of Software Projects

The class hierarchy of the aggregated ontology is shown in Fig. 3.14.

Instances of `Concept` are divided into instances of the classes `Project`, `Requirement`, `ActivityDiagram`, and `Element`. `Project` refers to the software project instantiated, while `Requirement` and `ActivityDiagram` are used to hold the requirements and diagrams of the static and the dynamic ontology, respectively. These instances ensure that the results of the ontology are traceable.

Any other ontology `Concept` is an `Element` of the project. Instances of type `Element` are divided into the subclasses `Resource`, `Activity`, `Action`, `Property`, and `Condition`. The `Resource` is the building block of any RESTful system, while `Action` is used to hold the CRUD actions performed on resources. `Activity` refers to an activity of the system (e.g., "create bookmark") that is connected to a `Resource` (e.g., "bookmark") and a CRUD `Action` (e.g., "create"). `Property` refers to a parameter required for a specific activity (e.g., "bookmark name" may be required for the "create bookmark"). Finally, an instance of `Condition` holds criteria that have to be met for an `Activity` to be executed. The properties of the ontology are shown in Table 3.4.

The properties including `Project`, `Requirement`, `ActivityDiagram`, and `Element` are used to ensure that any element of the diagram is traceable in the other two ontologies. The relations of the ontology classes are formed

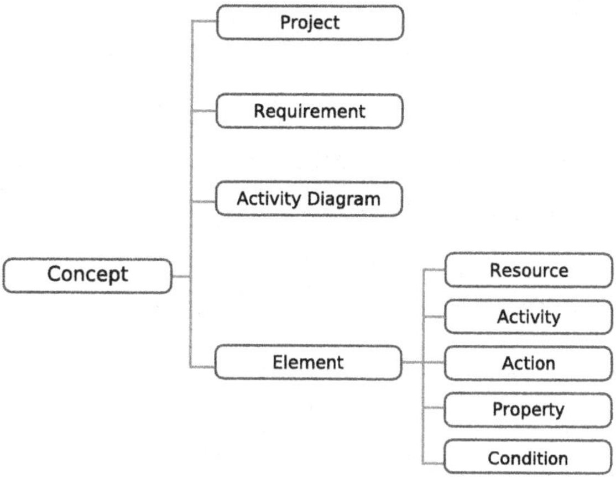

Fig. 3.14 Aggregated
ontology of software projects

Table 3.4 Properties of the aggregated ontology

OWL class	Property	OWL class
`Project`	`has_requirement`	`Requirement`
`Requirement`	`is_requirement_of`	`Project`
`Project`	`has_activity_diagram`	`ActivityDiagram`
`ActivityDiagram`	`is_activity_diagram_of`	`Project`
`Project`	`has_element`	`Element`
`Element`	`is_element_of`	`Project`
`Requirement/` `ActivityDiagram`	`contains_element`	`Element`
`Element`	`element_is_contained_in`	`Requirement/` `ActivityDiagram`
`Resource`	`has_activity`	`Activity`
`Activity`	`is_activity_of`	`Resource`
`Resource`	`has_property`	`Property`
`Property`	`is_property_of`	`Resource`
`Activity`	`has_action`	`Action`
`Action`	`is_action_of`	`Activity`
`Activity`	`has_condition`	`Condition`
`Property`	`is_condition_of`	`Activity`
`Activity`	`has_next_activity`	`Activity`
`Activity`	`has_previous_activity`	`Activity`

around two main subclasses of `Element`, `Resource`, and `Activity`. This
is quite expected since these two elements form the basis of a RESTful system.
Any system `Resource` may be connected to instances of type `Property` and
`Activity`, using has_property/is_property_of and has_activity/

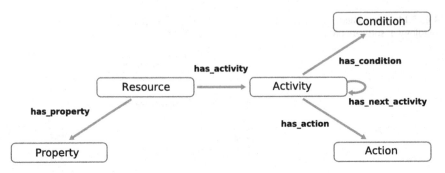

Fig. 3.15 Properties of the aggregated ontology

is_activity_of, respectively. The class Activity is connected to instances of type Action (via the properties has_action/is_action_of) and of type Condition (via the properties has_condition/is_condition_of), since it is necessary to keep track of the CRUD verbs to be used as well as any conditions that have to be met in order for the activity to be valid. Transitions are handled using the properties has_next_activity/has_previous_activity.

The properties are also visualized in Fig. 3.15 (excluding the properties relevant to Project, Requirement, and ActivityDiagram and including only one of the inverse properties for simplicity), where it is clear that Resource and Activity have central roles in the aggregated ontology.

The aggregated ontology is instantiated using the information provided by the static and dynamic ontologies of software projects. The static ontology contains several classes that refer to the static view of the system. Among them, we focus on actions performed on objects and any properties of these objects. In the static ontology, these elements are represented by the OWL classes OperationType, Object, and Property. Concerning the dynamic elements of a software system, the corresponding ontology covers not only actions, objects, and properties, but also the conditions of actions. The corresponding OWL classes are Action, Object, Property, and GuardCondition. Apart from the above classes, we also keep track of the Project that is instantiated, as well as the instances of type Requirement and ActivityDiagram derived from the static and dynamic ontologies, respectively. These three classes ensure that our ontologies are traceable and strongly linked to one another. The mapping of OWL classes from the static and dynamic ontologies to the aggregated ontology is shown in Table 3.5.

As shown in this table, instances of Requirement and ActivityDiagram are propagated to the aggregated ontology, while Project is used to ensure that the two ontology instantiations refer to the same project. Concerning the remaining classes of the aggregated ontology, several of them require merging the elements from the two ontologies. Thus, any Object of the static ontology and any Object of the dynamic ontology are added to the aggregated ontology one after another. If at any point an instance already exists in the aggregated ontology, then it is simply not

Table 3.5 Classes mapping from static and dynamic ontologies to aggregated ontology

OWL class of static ontology	OWL class of dynamic ontology	OWL class of aggregated ontology
`Project`	`Project`	`Project`
`Requirement`	–	`Requirement`
–	`ActivityDiagram`	`ActivityDiagram`
`OperationType`	`Action`	`Activity`
–	`GuardCondition`	`Condition`
`Object`	`Object`	`Resource`
`Property`	`Property`	`Property`

Table 3.6 Properties mapping from static and dynamic ontologies to aggregated ontology

OWL property of static ontology	OWL property of dynamic ontology	OWL property of aggregated ontology
`project_has`	–	`has_requirement`
`is_of_project`	–	`is_requirement_of`
–	`project_has_diagram`	`has_activity_diagram`
–	`is_diagram_of_project`	`is_activity_diagram_of`
`consists_of`	`diagram_has`	`contains_element`
`consist`	`is_of_diagram`	`element_is_contained_in`
`receives_action`	`is_object_of`	`has_activity`
`acts_on`	`has_object`	`is_activity_of`
`has_property`	`has_property`*	`has_property`
`is_property_of`	`is_property_of`*	`is_property_of`
–	`has_action`	`has_action`
–	`is_action_of`	`is_action_of`
–	`has_condition`*	`has_condition`
–	`is_condition_of`*	`is_condition_of`
–	`has_target`	`has_next_activity`
–	`has_source`	`has_previous_activity`

*derived property

added. However, any properties of this instance are also added (again if they do not exist); this ensures that the ontology is fully descriptive, yet without any redundant information. The mapping for OWL properties is shown in Table 3.6.

Note that some properties are not directly mapped among the ontologies. In such cases, properties can be derived from intermediate instances. For example, the aggregated ontology property `has_property` is directed from `Resource` to `Property`. In the case of the dynamic ontology, however, properties are connected to activities. Thus, for any `Activity`, e.g., "Create bookmark", we have to first find the respective `Object` ("bookmark") and then upon adding it to the

```
— !!Resource
  Name: String
  IsAlgorithmic: Boolean
  CRUDActivities: List of Create, Read, Update, and/or Delete
  Properties:
  — Name: String
    Type: Integer/Float/String/Boolean/null
    Unique: Boolean
    NamingProperty: Boolean
  — ...
  RelatedResources: List of String
```

Fig. 3.16 Schema of the YAML representation

aggregated ontology, we have to find the `Property` instances of the `Activity` (e.g., "bookmark name") and add them to the ontology along with the respective connection. This also holds for the condition properties `has_condition` and `is_condition_of`, which are instantiated using the instances of `GuardCondition` of the preceding `Transition`.

3.3.4.2 Ontology Software Artifacts to a YAML Representation

Upon instantiating the aggregated ontology, the next step is the transformation from the ontology to the specifications of the envisioned service. Specifications can be defined and stored in a variety of different formats, including Computationally Independent Models (CIMs), UML artifacts, etc. In our case, we design a representation in YAML,[14] which shall effectively describe the conceptual elements of the service under development and provide the user with the ability to modify (fine-grain) the model. After that, our YAML model can be used as a design artifact or even to directly produce a CIM, thus allowing developers to have a clear view of the specifications and in some cases even automate the construction of the service (e.g., as in [27]). YAML supports several well-known data structures, since it is designed to be easily mapped to programming languages. In our case, we use lists and associative arrays (i.e., key-value structures) to create a structure for resources, their properties, and the different types of information that have to be stored for each resource. The schema of our representation is shown in Fig. 3.16, where the main element is the RESTful resource (`Resource`).

A project consists of a list of resources. Several fields are defined for each resource, each with its own data type and allowed values. At first, each resource

---

[14]http://yaml.org/.

must have a name, which also has to be unique. Additionally, every resource may be either algorithmic (i.e., requiring some algorithm-related code to be written or some external service to be called) or non-algorithmic. This is represented using the `IsAlgorithmic` Boolean field. CRUD verbs/actions (or synonyms that can be translated to a CRUD verb) are usually not applied on algorithmic resources. There are four types of activities that can be applied to resources, in compliance with the CRUD actions (Create, Read, Update, and Delete). Each resource may support one or more of these actions, i.e., any combination of them, represented as a list (`CRUDActivities`).

Resources also have properties, which are defined as a list of objects. Each property has a `Name`, which is alphanumerical, as well as a `Type`, which corresponds to the common data types of programming languages, i.e., integers, floats, strings, and booleans. Furthermore, each property has two Boolean fields: `Unique` and `NamingProperty`. The former denotes whether the property has a unique value for each instance of the resource, while the latter denotes whether the resource is named after the value of this property. For example, a resource "user" could have the properties "username" and "email account". In this case, the "username" would possibly be unique, while "email account" could or could not be unique (e.g., a user may be allowed to declare more than one email accounts). Any instance of "user", however, should also be uniquely identified in the system. Thus, if we do not allow two users to have the same username, we could declare "username" as a naming property. Finally, each resource may have related resources. The field `RelatedResources` is a list of alphanumerical values corresponding to the names of other resources.

Extracting information from the aggregated ontology and creating the corresponding YAML file is a straightforward procedure. At first, instances of the OWL class `Resource` can directly be mapped to YAML objects of type `Resource`. Each resource is initially considered non-algorithmic. The flow of activities and conditions is used to find the types of verbs that are used on any resource as well as the related resources. Thus, for example, given an activity "Add bookmark" followed by an activity "Add tag", one may identify two resources, "bookmark" and "tag", where "tag" is also a related resource for "bookmark". Additionally, both "bookmark" and "tag" must have the "Create" CRUD activity enabled, since the verb "add" implies creating a new instance. The type for each verb is recognized using a lexicon. Whenever an action verb cannot be classified as any of the four CRUD types, a new algorithmic resource is created. Thus, for example, an activity "Search user" would spawn the new algorithmic resource "userSearch", and connecting it as a related resource of the resource "user". Finally, the properties of the resources are mapped to the `Properties` list field.

3.4 Case Study

In this section, we provide a case study for the example project Restmarks, which was introduced in Sect. 3.3.3. Consider Restmarks as a service that allows users to

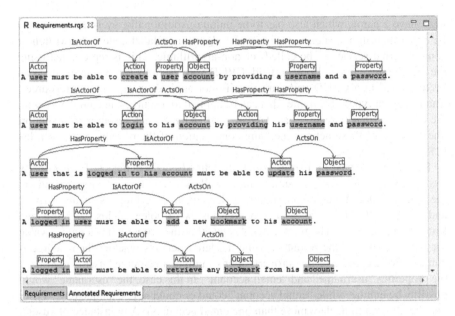

Fig. 3.17 Excerpt of the annotated functional requirements of project restmarks

store and retrieve online their bookmarks, share them with other users, and search for bookmarks by using tags. One could think of it as a social service for bookmarks. In the following paragraphs, we illustrate how one can use our system step by step, as in Fig. 3.4, to create such a product easily, while at the same time ensure that the produced service is fully functional and traceable.

The first step of the process includes entering and annotating functional requirements. An illustrative part of the requirements of Restmarks, annotated and refined using the Requirements Editor, is depicted in Fig. 3.17.

Upon adding the requirements, the user has to enter information about the dynamic view of the system. In this example, dynamic system representation is given in the form of storyboards. Let us assume Restmarks has the following dynamic scenarios:

1. Add Bookmark: The user adds a bookmark to his/her collection and optionally adds a tag to the newly added bookmark.
2. Create Account: The user creates a new account.
3. Delete Bookmark: The user deletes one of his/her bookmarks.
4. Login to Account: The user logs in to his/her account.
5. Search Bookmark by Tag System Wide: The user searches for bookmarks by giving the name of a tag. The search involves all public bookmarks.
6. Search Bookmark by Tag User Wide: The user searches for bookmarks by giving the name of a tag. The search involves the user's public and private bookmarks.
7. Show Bookmark: The system shows a specific bookmark to the user.
8. Update Bookmark: The user updates the information on one of his/her bookmarks.

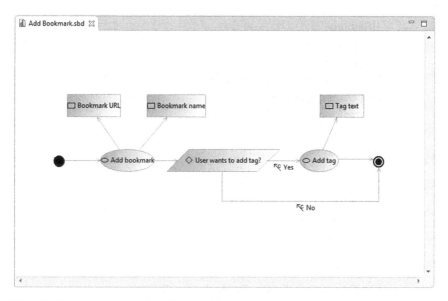

Fig. 3.18 Storyboard diagram "Add bookmark" of project restmarks

Table 3.7 Instantiated classes `Resource`, `Property`, and `Activity` for restmarks

OWL class	OWL instances
`Resource`	`bookmark, tag, account, password`
`Property`	`username, password, private, public, user`
`Activity`	`search_bookmark, add_tag, update_bookmark, delete_bookmark, mark_bookmark, retrieve_bookmark, delete_tag, add_bookmark, update_tag, update_password, login_account, create_account, get_bookmark`

The storyboards of the above scenarios are created using the Storyboard Creator. An example storyboard diagram using the tool is shown in Fig. 3.18. The storyboard of this figure refers to the first scenario of adding a new bookmark. The rest of the requirements and storyboards of the project are inserted accordingly.

Upon having composed of the requirements and storyboards, the next step involves creating the static and dynamic ontologies. The two ontologies are combined to provide the instances of the aggregated ontology. The instantiation of the classes `Resource`, `Property`, and `Activity` of the aggregated ontology is shown in Table 3.7.

Finally, the YAML representation that describes the project is exported from the aggregated ontology. For Restmarks, the YAML file is shown in Fig. 3.19.

Once the YAML representation of the envisioned system is produced, the developer has complete and traceable specifications that can be subsequently used for various purposes, including requirements validation, system development, and soft-

```
—  !!Resource
   Name: account
   IsAlgorithmic: false
   CRUDActivities: [Create, Read, Update, Delete]
   Properties: [username, password]
   RelatedResources: [bookmark]
—  !!Resource
   Name: tagSearch
   IsAlgorithmic: true
   CRUDActivities: []
   Properties: []
   RelatedResources: []
—  !!Resource
   Name: tag
   IsAlgorithmic: false
   CRUDActivities: [Create, Read, Update, Delete]
   Properties: [name, description]
   RelatedResources: [tagSearch]
—  !!Resource
   Name: bookmark
   IsAlgorithmic: false
   CRUDActivities: [Create, Read, Update, Delete]
   Properties: [url, scope]
   RelatedResources: [tag]
```

Fig. 3.19 Example YAML file for project restmarks

ware reuse. Upon demonstrating how one can instantiate our models, the tasks of requirements mining and reuse are illustrated in the following chapter, while for more information about the potential model-to-code transformation the interested reader is referred to [27].

3.5 Conclusion

Lately, the problem of automating the requirements elicitation and specification extraction process has attracted the attention of several researchers. However, current approaches do not support different types of requirements and usually depend on domain-specific heuristics and/or controlled languages. In this work, we have designed a methodology that allows developers to model their envisioned system using software requirements from multimodal formats. Our system receives input in the form of functional requirements and action flows in the form of storyboards, thus covering both the structural and the behavioral views of the envisioned system. The use of NLP and semantics facilitates the extraction of specifications from these requirements, while the designed ontologies produce a traceable model.

To further illustrate the advantages of our approach, we focus on RESTful web service development and thus illustrate how the specifications of a service can be easily produced. The produced model (YAML file) comprises all the required elements of the service, including resources, CRUD actions and properties, and support for hypermedia links. Future work on our methodology may lie in several directions. The continuous improvement of the tools by receiving feedback from users is in our immediate plans, while future research includes further assessing our methodology in industrial settings.

References

1. van Lamsweerde A (2009) Requirements engineering: from system goals to UML models to software specifications. Wiley, Hoboken
2. Luisa M, Mariangela F, Pierluigi NI (2004) Market research for requirements analysis using linguistic tools. Requir Eng 9(1):40–56
3. Boehm B, Basili VR (2001) Software defect reduction top 10 list. Computer 34:135–137
4. Kaindl H, Smialek M, Svetinovic D, Ambroziewicz A, Bojarski J, Nowakowski W, Straszak T, Schwarz H, Bildhauer D, Brogan JP, Mukasa KS, Wolter K, Krebs T (2007) Requirements specification language definition: defining the ReDSeeDS languages, Deliverable D2.4.1. Public deliverable, ReDSeeDS (Requirements driven software development system) project
5. Smialek M. (2012) Facilitating transition from requirements to code with the ReDSeeDS tool. In: Proceedings of the 2012 IEEE 20th international requirements engineering conference (RE), RE '12. IEEE Computer Society, Washington, pp 321–322
6. Wynne M, Hellesoy A (2012) The cucumber book: behaviour-driven development for testers and developers. Pragmatic bookshelf
7. North D (2003) JBehave: a framework for behaviour driven development. http://jbehave.org/. Accessed March 2013
8. Mylopoulos J, Castro J, Kolp M (2000) Tropos: a framework for requirements-driven software development. Information systems engineering: state of the art and research themes. Springer, Berlin, pp 261–273
9. Yu ES-K (1995) Modelling strategic relationships for process reengineering. PhD thesis, University of Toronto, Toronto, Ont., Canada. AAINN02887
10. Abbott RJ (1983) Program design by informal english descriptions. Commun ACM 26(11):882–894
11. Booch G (1986) Object-oriented development. IEEE Trans Softw Eng 12(1):211–221
12. Saeki M, Horai H, Enomoto H (1989) Software development process from natural language specification. In: Proceedings of the 11th international conference on software engineering, ICSE '89. ACM, New York, pp 64–73,
13. Mich L (1996) NL-OOPS: from natural language to object oriented requirements using the natural language processing system LOLITA. Nat Lang Eng 2(2):161–187
14. Harmain HM, Gaizauskas R (2003) CM-Builder: a natural language-based CASE tool for object-oriented analysis. Autom Softw Eng 10(2):157–181
15. Casta v, Ballejos L, Caliusco ML, Galli MR (2010) The use of ontologies in requirements engineering. Glob J Res Eng 10(6)

16. Siegemund K, Thomas EJ, Zhao Y, Pan J, Assmann U (2011) Towards ontology-driven requirements engineering. In: Workshop semantic web enabled software engineering at 10th international semantic web conference (ISWC), Bonn

17. Happel H-J, Seedorf S (2006) Applications of ontologies in software engineering. In: Proceedings of the 2nd international workshop on semantic web enabled software engineering (SWESE 2006), held at the 5th international semantic web conference (ISWC 2006), pp 5–9

18. Dermeval D, Vilela J, Bittencourt I, Castro J, Isotani S, Brito P, Silva A (2015) Applications of ontologies in requirements engineering: a systematic review of the literature. Requir Eng 1–33

19. Kambhatla N (2004) Combining lexical, syntactic, and semantic features with maximum entropy models for extracting relations. In: Proceedings of the ACL 2004 on interactive poster and demonstration sessions, ACLdemo'04. ACL, Stroudsburg, pp 178–181

20. Zhao S, Grishman R (2005) Extracting relations with integrated information using kernel methods. In: Proceedings of the 43rd annual meeting on association for computational linguistics, ACL '05. Association for Computational Linguistics, Stroudsburg, pp 419–426

21. GuoDong Z, Jian S, Jie Z, Min Z (2005) Exploring various knowledge in relation extraction. In: Proceedings of the 43rd annual meeting on association for computational linguistics, ACL '05. Association for Computational Linguistics, Stroudsburg, pp 427–434

22. Bunescu R, Mooney RJ (2005) Subsequence kernels for relation extraction. In: Advances in Neural Information Processing Systems, vol. 18: proceedings of the 2005 conference (NIPS), pp 171–178

23. Zelenko D, Aone C, Richardella A (2003) Kernel methods for relation extraction. J Mach Learn Res 3:1083–1106

24. Culotta A, Sorensen J (2004) Dependency tree kernels for relation extraction. In: Proceedings of the 42nd annual meeting on association for computational linguistics, ACL '04. Association for Computational Linguistics, Stroudsburg, pp 423–429

25. Bunescu RC, Mooney RJ (2005) A shortest path dependency kernel for relation extraction. In: Proceedings of the conference on human language technology and empirical methods in natural language processing. Association for Computational Linguistics, Stroudsburg, pp 724–731

26. Bach N, Badaskar S (2007) A review of relation extraction. Carnegie Mellon University, Language Technologies Institute

27. Zolotas C, Diamantopoulos T, Chatzidimitriou K, Symeonidis A (2016) From requirements to source code: a model-driven engineering approach for RESTful web services. Autom Softw Eng 1–48

28. Roth M, Diamantopoulos T, Klein E, Symeonidis AL (2014) Software requirements: a new domain for semantic parsers. In: Proceedings of the ACL 2014 workshop on semantic parsing, SP14, Baltimore, pp 50–54

29. Diamantopoulos T, Roth M, Symeonidis A, Klein E (2017) Software requirements as an application domain for natural language processing. Lang Resour Eval 51(2):495–524

30. Björkelund A, Hafdell L, Nugues P (2009) Multilingual semantic role labeling. In: Proceedings of the 13th conference on computational natural language learning: shared task, CoNLL '09. Association for Computational Linguistics, Stroudsburg, pp 43–48

31. Bohnet B (2010) Top accuracy and fast dependency parsing is not a contradiction. In: Proceedings of the 23rd international conference on computational linguistics, Beijing, pp 89–97

32. Fan RE, Chang KW, Hsieh CJ, Wang XR, Lin CJ (2008) LIBLINEAR: a library for large linear classification. J Mach Learn Res 9:1871–1874

Chapter 4
Mining Software Requirements

4.1 Overview

Having proposed a model for transforming and storing functional requirements in the previous chapter, in this chapter, we focus on the problem of requirements identification and co-occurrence. It is common knowledge that inaccurate, incomplete, or undefined requirements have been found to be the most common reason of failure for software projects [1]. Furthermore, continuous changes to initial requirements can lead to faults [1], while reengineering costs as a result of poorly specified requirements are considerably high [2]. In this context, identifying the proper requirements for the software project at hand is extremely important.

In this chapter, we view the challenge of requirements identification under the prism of software reuse. As already mentioned, the introduction of the open-source software initiatives and the component-based nature of software has led to a new way of developing software by relying more and more on reusing components in a rapid prototyping context. The need for reusing existing components has become more eminent than ever, since component reuse can reduce the time and effort spent during all phases of the software development life cycle, including requirements elicitation, specification extraction and modeling, source code writing, and software maintenance/testing. As a result, there are several approaches toward applying data mining techniques to recommend software components that address the required functionality and are of high quality. Most of these approaches, however, focus on the source code of the components [3–5], on quality information/metrics [6], and on information from online repositories [7, 8]. Requirements reuse cannot be easily tackled as requirements are usually expressed in natural language text and UML models, which are often ambiguous or incomprehensible to developers and stakeholders [9].

The benefits from requirements reuse are evident regardless of the type of software product built and the software development methodology adopted. Software is built based on requirements identified, irrespective of whether they have been collected all at once, iteratively and/or through throw-away prototypes. What is, thus,

© Springer Nature Switzerland AG 2020
T. Diamantopoulos and A. L. Symeonidis, *Mining Software Engineering Data for Software Reuse*, Advanced Information and Knowledge Processing,
https://doi.org/10.1007/978-3-030-30106-4_4

important is to *design a model capable of storing software requirements* and *develop a methodology that will enable requirements reuse*. The model we propose is the one we designed in the previous chapter, which allows seamless instantiation or even migration of existing requirements, and supports reuse regardless of the software engineering methodology applied. The mining methodology is another important challenge that is obviously highly dependent on the employed model.

As far as requirements quality check is concerned, several research efforts involve storing software requirements in formal models [10–13], which allow requirements engineers to have full control of the system design and detect errors at an early stage of the project life cycle, which is usually much more cost-effective than finding and fixing them at a later stage [14]. Concerning functional requirements elicitation, most approaches involve constructing and instantiating models using domain-specific vocabularies. The models can be subsequently used to perform validation [15, 16] or to recommend new requirements [17, 18] by identifying missing entities and relations [19]. Other lines of research include identifying dependencies among requirements and distributing them according to their relevance to stakeholders [20], as well as recovering traceability links among requirements and software artifacts and utilizing them to identify potentially changed requirements [21]. Though effective, most of these efforts are confined to specific domains, mainly due to the lack of annotated requirements models for multiple domains. UML diagram/model mining techniques suffer more or less from the same issues as functional requirements approaches. Most semantics-enabled methods [22, 23] are based on the existence of domain-specific information. On the other hand, domain-agnostic techniques [24–26] can be greatly facilitated by appropriate data handling of requirements models, as their shortcomings usually lie on incorporating structure or flow information, e.g., for use case or activity diagrams.

In this chapter, we employ the ontology models designed in the previous chapter for storing functional requirements and UML diagrams and develop a mining methodology to support requirements reuse. We employ data mining techniques in two reuse axes: functional requirements and UML models. Concerning functional requirements, association rule mining techniques and heuristics are used to determine whether the requirements of a software project are complete and recommend new requirements (as discussed in [27]). Concerning UML models, matching techniques are used in order to find similar diagrams and thus allow the requirements engineer to improve the existing functionality and the data flow/business flow of his/her project.

4.2 State of the Art on Requirements Mining

4.2.1 Functional Requirements Mining

Early research efforts in recommendation systems for requirements elicitation were focused on domain analysis and thus used linguistics (vocabularies, lexicons) to

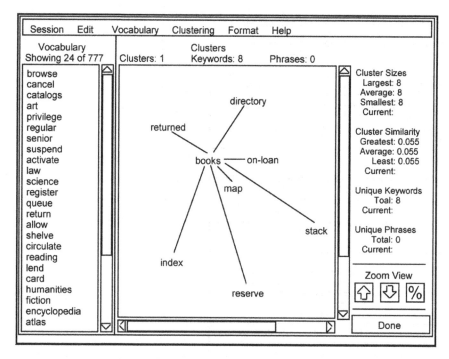

Fig. 4.1 Cluster editor screen of DARE [19]

determine project domains and identify missing entities and relations at requirements' level. DARE [19] is such a tool that utilizes multiple sources of information, including the requirements, the architecture, and the source code of a project. DARE extracts entities and relations from several projects and then uses clustering to identify common entities and thus recommend similar artifacts and architectural schemata for each project. Figure 4.1 depicts an example mockup screen of the tool for clustering terms relevant to a library management domain. Kumar et al. [15], on the other hand, make use of ontologies in order to store requirements and project domains and extract software specifications. Ghaisas and Ajmeri [16] further develop a Knowledge-Assisted Ontology-Based Requirements Evolution (K-RE) repository in order to facilitate requirements elicitation and resolve any conflicts between change requests.

There are also several approaches that aspire to identify the features of a system through its requirements and recommend new ones. Chen et al. [17] used requirements from several projects and constructed relationship graphs among requirements. The authors then employed clustering techniques to extract domain information. Their system can identify features that can be used to create a feature model of projects that belong to the same domain. Similar work was performed by Alves et al. [18], who employed the vector space model to construct a domain feature model and used latent semantic analysis to find similar requirements by clustering

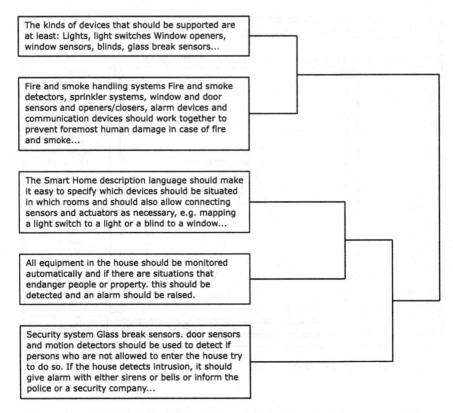

Fig. 4.2 Example clustering of requirements for a smart home system [18]

them into domains. An example for this line of work is shown in Fig. 4.2, which depicts the hierarchical clustering of five textual requirements extracted from a document describing a smart home system. The first two requirements (from top to bottom) are quite similar as they are both related to devices and their integration. The third requirement has similar terms with the fourth one, while the fifth one is also closer to the fourth one, as both requirements describe actions for detecting danger. On the other hand, Dumitru et al. [28] employed association rule mining to analyze requirements and subsequently cluster them into feature groups. These groups can be used to find similar projects or to recommend a new feature for a project.

From a different perspective, there are also approaches that explore the relation between requirements and stakeholders [29–31]; the related systems are given as input ratings for the requirements of individual stakeholders and use collaborative filtering [32] to provide recommendations and prioritize requirements according to stakeholder preferences. However, these approaches, and any approach focusing on non-functional requirements [33], deviate from the scope of this work.

Based on the above discussion, one easily notices that most approaches are domain-centered [15–19], which is actually expected as domain-specific information can improve the understanding of the software project under analysis. Still, as noted by Dumitru et al. [28], these domain analysis techniques are often not applicable due to the lack of annotated requirements for a specific domain. And although feature-based techniques [28, 33] do not face such issues, their scope is too high level and they are not applied on fine-grained requirements.

In the context of our work, we build a system that focuses on functional requirements and provides low-level recommendations, maintaining a semantic domain-agnostic outlook. To do so, we employ the static ontology and the semantic role labeler of the previous chapter to extract actors, actions, objects, and properties from functional requirements, and effectively index them. The descriptive power of our model allows us to recommend requirements in a comprehensive and domain-agnostic manner.

4.2.2 UML Models Mining

Early efforts for UML model mining employed information retrieval techniques and were directed toward use case scenarios [24, 25]. Typically, such scenarios can be defined as sets of *events* triggered by *actions* of *actors*, and they also include *authors* and *goals*. Thus, research on scenario reuse is mainly directed toward representing event flows and comparing them to find similar use cases [25]. Other approaches involve representing UML diagrams to graphs and detecting similar graphs (diagrams) using graph matching techniques. In such graph representations, the vertices comprise the object elements of UML use case, class, and sequence diagrams, while the edges denote the associations among these elements. The employed graph matching techniques can either be exact [34–36] or inexact [37]. Similar work by Park and Bae [38] involves using Message Object Order Graphs (MOOGs) to store sequence diagrams, where the nodes and the edges represent the messages and the flow among them.

Despite the effectiveness of graph-based methods under certain scenarios (e.g., structured models), applying them to UML diagrams lacks semantics. As noted by Kelter et al. [26], UML model mining approaches should focus on creating a semantic model, rather than arbitrarily applying similarity metrics. For this reason, UML diagrams (and XML structures) are often represented as ordered trees [26, 39, 40]. The main purpose of these efforts is to first design a data model that captures the structure of UML diagrams and then apply ordered tree similarity metrics. Indicatively, the model of Kelter et al. [26] is shown in Fig. 4.3. It consists of the following elements: a Document, a set of Elements, each one with its type (ElementType), a set of Attributes for each Element, and a set of References between elements. The model supports several types of data, including not only UML models, but also more generic ones, such as XML documents. Furthermore, the Elements have hierarchy, which is quite useful for UML Class diagrams, where examples of Elements would include packages, classes, methods, etc., which are indeed hierarchical (e.g., packages include classes).

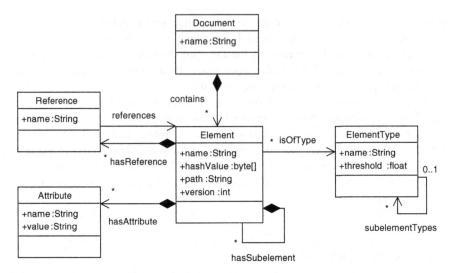

Fig. 4.3 Data model of the UML differencing technique of Kelter et al. [26]

These approaches are effective for static diagram types (use case, class); however, they cannot efficiently represent data flows or action flows, and thus they do not adhere to dynamic models (activity, sequence). Furthermore, they usually employ string difference techniques, and thus they are not applicable to scenarios with multiple sources of diagrams. To confront this challenge, researchers have attempted to incorporate semantics through the use of domain-specific ontologies [22, 23, 41]. Matching diagram elements (actors, use cases, etc.) according to their semantic distance has indeed proven effective as long as domain knowledge is available; most of the time, however, domain-specific information is limited.

We focus on the requirements elicitation phase and propose two mining methodologies, one for use case diagrams and one for activity diagrams. For use case diagram matching, we handle use cases as discrete nodes and employ semantic similarity metrics, thus combining the advantages of graph-based and information retrieval techniques and avoiding their potential limitations. For activity diagram matching, we construct a model that represents activity diagrams as sequences of action flows. This way, the dynamic nature of the diagrams is modeled effectively, without using graph-based methods that could result to over-engineering. Our algorithm is similar to ordered tree methods [26, 39, 40], which are actually a golden mean between unstructured (e.g., Information Retrieval) and heavily structured (e.g., graph-based) methods. Our matching methodologies further use a semantic scheme for comparing strings, which is, however, domain-agnostic and thus can be used in scenarios with diagrams originating from various projects.

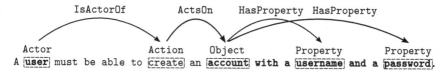

Fig. 4.4 Example annotated requirement

4.3 Functional Requirements Mining

As already mentioned, the modeling part of our functional requirements mining methodology comprises the static ontology (and obviously all supporting tools, such as the NLP parser, etc.). Thus, for the mining part, we assume that the engineer has already instantiated models of multiple projects using our tools. To design and evaluate our requirements mining model, we use a diverse dataset of 30 projects, which includes student projects, industrial prototypes, as well as RESTful prototype applications. In total, these projects have 514 functional requirements and amount to 7234 entities and 6626 relations among them. An example annotated requirement is shown in Fig. 4.4.

Our system can relate domain-agnostic terms between different projects. At first, the entities and relations for each requirement (and thus for each project) are extracted. For instance, for the requirement of Fig. 4.4, we may extract the terms `user`, `create`, `account`, `username`, and `password`, as well as the corresponding relations among them: `user_IsActorOf_create`, `create_ActsOn_account`, `account_HasProperty_username`, and `account_HasProperty_password`. Relating two requirements (or two projects in general) requires semantically relating their terms. For instance, if we have another project where each user also has an account, yet the chosen term is `profile`, then these two terms have to be marked as semantically similar.

In order to mark semantically similar terms, we require an index of words and a similarity measure. We use WordNet [42] as our index and employ the MIT Java Wordnet Interface [43] and the Java Wordnet::Similarity library[1] to interface with it. There are several methods for computing the similarity between two terms [44]. However, most of them either do not employ semantics or they are not correlated to human judgment. As a result, we use the information content measure introduced by Lin [45], which conforms with human judgment more often than other metrics. We define the similarity between two terms (i.e., WordNet classes) C_1 and C_2 as follows:

$$sim(C_1, C_2) = \frac{2 \cdot \log P(C_0)}{\log P(C_1) + \log P(C_2)} \qquad (4.1)$$

---

[1]http://users.sussex.ac.uk/~drh21/.

Fig. 4.5 Example WordNet
fragment where `record` is
the most specific class of
`account` and `profile`

```
                              entity
                                |
                           abstraction
                                |
                          communication
                                |
                            indication
                                |
                             evidence
                                |
                              record
                            ⌢⌢⌢⌢⌢
                    account     history
                                   |
                               biography
                                   |
                                profile
```

where C_0 is the most *specific* class that contains both C_1 and C_2. For example, the most specific class that describes the terms `account` and `profile` is the class `record`, as shown also in the relevant WordNet fragment of Fig. 4.5.

Upon having found C_0, we compute the similarity between the two terms using Eq. (4.1), which requires the *information content* for each of the three WordNet classes. The information content of a WordNet class is defined as the log probability that a corpus term belongs in that class.[2] Thus, for instance, `record`, `account`, and `profile` have information content values equal to 7.874, 7.874, and 11.766, respectively, and the similarity between `account` and `profile` is $2 \cdot 7.874/(7.874 + 11.766) = 0.802$. Finally, two terms are assumed to be semantically similar if their similarity value (as determined by Eq. (4.1)) is higher than a threshold t. We set t to 0.5, and thus we now have 1512 terms for the 30 projects, out of which 1162 are distinct.

Having performed this type of analysis, we now have a dataset with one set of *items* per software project and can extract useful *association rules* using association rule mining [46]. Let $P = \{p_1, p_2, \ldots, p_m\}$ be the set of m software projects and $I = \{i_1, i_2, \ldots, i_n\}$ be the set of all n items. *Itemsets* are defined as subsets of I. Given an itemset X, its support is defined as the number of projects in which all of its items appear in

$$\sigma(X) = |\{p_i | X \subset p_i, p_i \in P\}| \tag{4.2}$$

Association rules are expressed in the form $X \rightarrow Y$, where X and Y are disjoint itemsets. An example rule that can be extracted from the items of the requirement of Fig. 4.4 is $\{account\_HasProperty\_username\} \rightarrow \{account\_HasProperty\_password\}$.

The two metrics used to determine the strength of a rule are its *support* and its *confidence*. Given an association rule $X \rightarrow Y$, its support is the number of projects

---

[2]We used the precomputed information content data of the Perl WordNet similarity library [44], available at http://www.d.umn.edu/~tpederse/.

Table 4.1 Sample association rules extracted from the dataset

No	Association rule	Support (σ)	Confidence (c)
1	$provide\_Acts On\_product \rightarrow$ $system\_Is Actor Of\_provide$	0.167	1.0
2	$system\_Is Actor Of\_validate \rightarrow$ $user\_Is Actor Of\_login$	0.1	1.0
3	$user\_Is Actor Of\_buy \rightarrow$ $system\_Is Actor Of\_provide$	0.1	1.0
4	$administrator\_Is Actor Of\_add \rightarrow$ $administrator\_Is Actor Of\_delete$	0.167	0.833
5	$user\_Is Actor Of\_logout \rightarrow$ $user\_Is Actor Of\_login$	0.167	0.833
6	$user\_Is Actor Of\_add \rightarrow$ $user\_Is Actor Of\_delete$	0.133	0.8
7	$user\_Is Actor Of\_access \rightarrow$ $user\_Is Actor Of\_view$	0.1	0.75
8	$edit\_Acts On\_product \rightarrow$ $add\_Acts On\_product$	0.1	0.75
9	$administrator\_Is Actor Of\_delete \rightarrow$ $administrator\_Is Actor Of\_add$	0.167	0.714
10	$user\_Has Property\_contact \rightarrow$ $user\_Is Actor Of\_search$	0.133	0.5

for which the rule is applicable, and it is given as

$$\sigma(X \rightarrow Y) = \frac{\sigma(X \cup Y)}{|P|} \qquad (4.3)$$

The confidence of the rule indicates how frequently items in Y appear in X, and it is given as

$$c(X \rightarrow Y) = \frac{\sigma(X \cup Y)}{\sigma(X)} \qquad (4.4)$$

We use the Apriori association rule mining algorithm [47] in order to extract association rules with support and confidence above certain thresholds. For our dataset, we set the minimum support to 0.1, so that any rule has to be contained in at least 10% of the projects. We also set the minimum confidence to 0.5, so that the extracted rules are confirmed at least half of the time that their antecedents are found. The execution of Apriori resulted in 1372 rules, a fragment of which is shown in Table 4.1.

Several of these rules are expected. For example, rule 2 indicates that in order for a user to login, the system must first validate the account. Also, rule 5 implies that logout functionality should co-occur in a system with login functionality.

Table 4.2 Activated rule heuristics for a software project

Antecedent	Consequent	Conditions	Result
$[Actor1, Action1]$	$[Actor2, Action2]$	$Actor2 \in p$	$[Actor2, Action2, Object]$, $\forall Object \in p$: $[Actor1, Action1, Object]$
$[Action1, Object1]$	$[Actor2, Action2]$	$Actor2 \in p$	$[Actor2, Action2, Object1]$
$[Actor1, Action1]$	$[Action2, Object2]$	$Object2 \in p$	$[Actor1, Action2, Object2]$
$[Action1, Object1]$	$[Action2, Object2]$	$Object2 \in p$	$[Actor, Action2, Object2]$, $\forall Actor \in p$: $[Actor, Action1, Object1]$
* (except for the above)	$[Action2, Object2]$	$Object2 \in p$	$[Actor, Action2, Object2]$, $\forall Actor, Action \in p$: $[Actor, Action, Object2]$
*	$[Any2, Property2]$	$Any2 \in p$	$[Any2, Property2]$

Upon having extracted the rules, we use them to recommend new requirements for software projects. At first, we define a new set of items p for the newly added software project. Given this set of items and the rules, we extract the *activated* rules R. A rule $X \rightarrow Y$ is activated for the project with set of items p if all items present in X are also contained in p (i.e., $X \subset p$). The set of activated rules R is then *flattened* by creating a new rule for each combination of antecedents and consequents of the original rule, so that the new rules contain single items as antecedents and consequents. Given, e.g., the rule $X \rightarrow Y$ where the itemsets X and Y contain the items $\{i_1, i_2, i_3\}$ and $\{i_4, i_5\}$, respectively, the new flattened rules are $i_1 \rightarrow i_4$, $i_1 \rightarrow i_5$, $i_2 \rightarrow i_4$, $i_2 \rightarrow i_5$, $i_3 \rightarrow i_4$, and $i_3 \rightarrow i_5$. We also propagate the support and the confidence of the original rules to these new flattened rules, so that they are used as important criteria. Finally, given the set of items of a project p and the flattened activated rules for this project, our system provides recommendations of new requirements using the heuristics of Table 4.2.

Concerning the heuristics for consequent $[Actor2, Action2]$, which corresponds to an `Actor2_IsActorOf_Action2` item, the recommended requirement includes the actor and the action of the consequent as well as an $Object$ that is determined by the antecedent. Given, e.g., an antecedent $[create, bookmark]$ and a consequent $[user, edit]$, the new recommended requirement will be $[user, edit, bookmark]$. Concerning the consequent $[Action2, Object2]$, which corresponds to an `Action2_ActsOn_Object2` item, the recommended requirement includes the action and the object of the consequent as well as the actor that is determined by the antecedent. Given, e.g., an antecedent $[user, profile]$ and a consequent $[create, profile]$, the new recommended requirement will be $[user, create, profile]$. Finally, any rule with a `HasProperty` consequent (and any antecedent) leads to new recommended requirements of the form $[Any, Property]$. An example requirement would be $[user, profile]$. Using the static ontology,

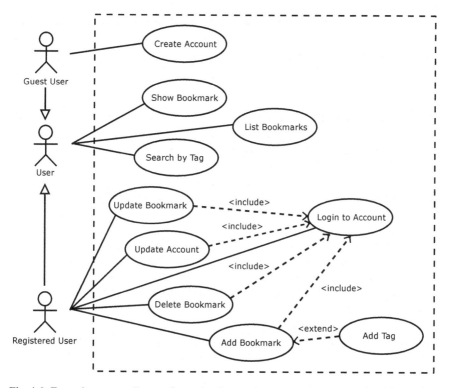

Fig. 4.6 Example use case diagram for project Restmarks

we are also able to reconstruct requirements, in the formats "The *Actor* must be able to *Action Object*" and "The *Any* must have *Property*".

4.4 UML Models Mining

To design our UML diagrams mining model, we use a dataset of use case and activity diagrams, which includes the diagrams of project Restmarks, the demo social service that was used in the previous chapter. An example use case diagram from Restmarks that contains 10 use cases and 3 actors is shown in Fig. 4.6.

As use cases refer to static information, they are stored in the static ontology with a model that is mostly flat. Specifically, the model comprises the actors and the use cases of the diagram. For example, the diagram D of Fig. 4.6 has a model that consists of two sets: the set of actors $A = \{User, Registered User, Guest User\}$ and the set of use cases $UC = \{Add Bookmark, Update Bookmark, Update Account, Show Bookmark, Search by Tag, Add Tag, Login to Account, Delete Bookmark\}$. Given two diagrams D_1 and D_2, our matching scheme involves two sets for each diagram: one set for the actors A_1 and A_2, respectively, and one set for the use cases UC_1 and UC_2, respectively. The similarity between the diagrams is computed by the following equation:

$$s(D_1, D_2) = \alpha \cdot s(A_1, A_2) + (1 - \alpha) \cdot s(UC_1, UC_2) \qquad (4.5)$$

where s denotes the similarity between two sets (either of actors or of use cases) and α denotes the importance of the similarity of actors for the diagrams. α was set to the proportion of the number of actors divided by the number of use cases of the queried diagram. Given, e.g., a diagram with 3 actors and 10 use cases, α is set to 0.3.

The similarity between two sets, either actors or use cases, is given by the combination between all the matched elements with the highest score. Given, e.g., two sets $\{user, administrator, guest\}$ and $\{administrator, user\}$, the best possible combination is $\{(user, user), (administrator, administrator), (guest, null)\}$, and the matching would return a score of $2/3 = 0.66$. We employ the semantic measure of the previous section to provide a similarity score between strings. Given two strings S_1 and S_2, we first split them into tokens, i.e., $tokens(S_1) = \{t_1, t_2\}$, $tokens(S_2) = \{t_3, t_4\}$, and then determine the combination of tokens with the maximum token similarity scores. The final similarity score between the strings is determined by averaging over all tokens. For example, given the strings "Get bookmark" and "Retrieve bookmarks", the best combination is ("get", "retrieve") and ("bookmark", "bookmarks"). Since the semantic similarity between "get" and "retrieve" is 0.677, and the similarity of "bookmark" with "bookmarks" is 1.0, the similarity between the strings is $(0.677 + 1)/2 = 0.8385$.

Given, for example, the diagram of Fig. 4.6 and the diagram of Fig. 4.7, the matching between the diagram elements is shown in Table 4.3, while the final score (using Eq. (4.5)) is 0.457.

Concerning recommendations, the engineer of the second diagram could consider adding a guest user. Furthermore, he/she could consider adding use cases for listing or updating bookmarks, adding tags to bookmarks, or updating account data.

Concerning activity diagrams, we require a representation that would view the diagram as a flow model. An example diagram of Restmarks is shown in Fig. 4.8.

Activity diagrams consist mostly of sequences of activities and possible conditions. Concerning conditions (and forks/joins), we split the flow of the diagram. Hence, an activity diagram is actually treated as a set of sequences, each of which involves the activities required to traverse the diagram from its start node to its end node. For instance, the diagram of Fig. 4.8 spawns a sequence $StartNode > Logged In? > Login to account > Provide bookmark URL > Create Bookmark > Add tag > User wants to add tag? > EndNode$. Upon having parsed two diagrams and having extracted one set of sequences per diagram, we compare the two sets. In this case and in contrast with use case diagram matching, we set a threshold t_{ACT} for the semantic string similarity metric. Thus, two strings are considered similar if their similarity score is larger than this threshold. We set t_{ACT} to 0.5.

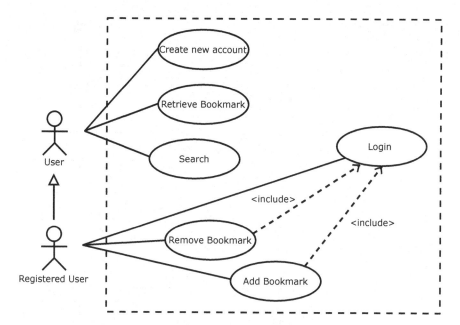

Fig. 4.7 Example use case diagram for matching with the one of Fig. 4.6

Table 4.3 Matching between the diagrams of Figs. 4.6 and 4.7

Diagram 1	Diagram 2	Score
User	User	1.00
Registered user	Registered user	1.00
Guest user	null	0.00
Delete bookmark	Remove bookmark	0.86
Show bookmark	Retrieve bookmark	0.50
Add bookmark	Add bookmark	1.00
Create account	Create new account	0.66
Search by tag	Search	0.33
Login to account	Login	0.33
List bookmarks	null	0.00
Update bookmark	null	0.00
Update account	null	0.00
Add tag	null	0.00

We use the *Longest Common Subsequence (LCS)* [48] in order to determine the similarity score between two sequences. The LCS of two sequences S_1 and S_2 is defined as the longest subsequence of common (yet not consecutive) elements between them. Given, e.g., the sequences $[A, B, D, E, G]$ and $[A, B, E, H]$, their

Fig. 4.8 Example activity
diagram for project
Restmarks

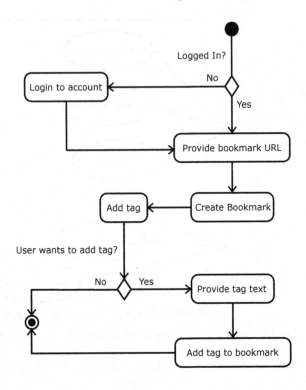

LCS is $[A, B, E]$. Finally, the similarity score between the sequences is defined as

$$sim(S_1, S_2) = 2 \cdot \frac{|LCS(S_1, S_2)|}{|S_1| + |S_2|} \qquad (4.6)$$

The similarity score is normalized in $[0, 1]$. Given that we have two sets of sequences (one for each of the two diagrams), their similarity is given by the best possible combination between the sequences, i.e., the combination that results in the highest score. For example, given two sets $\{[A, B, E], [A, B, D, E], [A, B, C, E]\}$ and $\{[A, B, E], [A, C, E]\}$, the combination that produces the highest possible score is $\{([A, B, E], [A, B, E]), ([A, B, C, E], [A, C, E]), ([A, B, D, E], null)\}$. Finally, the similarity score between the diagrams is the average of their LCS scores.

For example, let us consider matching the diagrams of Figs. 4.8 and 4.9. Our system returns the matching between the sequences of the diagrams, as shown in Table 4.4, while the total score, which is computed as the mean of these scores, is $(0.833 + 0.714 + 0 + 0)/4 = 0.387$.

The matching process between the sequences indicates that the engineer of the second diagram could add a new flow that would include the option to add a tag to the newly created bookmark.

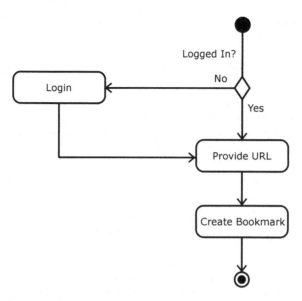

Fig. 4.9 Example activity diagram for matching with the one of Fig. 4.8

Table 4.4 Matching between the diagrams of Figs. 4.8 and 4.9

Diagram 1	Diagram 2	Score
StartNode > Logged In? > Provide bookmark URL > Create Bookmark > Add tag > User wants to add tag? > EndNode	StartNode > Logged In? > Provide URL > Create Bookmark > EndNode	0.833
StartNode > Logged In? > Login to account > Provide bookmark URL > Create Bookmark > Add tag > User wants to add tag? > EndNode	StartNode > Logged In? > Login > Provide URL > Create Bookmark > EndNode	0.714
StartNode > Logged In? > Provide bookmark URL > Create Bookmark > Add tag > User wants to add tag? > Provide tag text > Add tag to bookmark > EndNode	null	0.000
StartNode > Logged In? > Login to account > Provide bookmark URL > Create Bookmark > Add tag > User wants to add tag? > Provide tag text > Add tag to bookmark > EndNode	null	0.000

4.5 Evaluation

4.5.1 Functional Requirements Mining

In order to demonstrate the validity of our approach, we have performed requirements mining and provide recommendations for a small-scale software project. The selected project is project Restmarks, the demo social service for bookmarks that was introduced in the previous chapter. The requirements of the service are shown in Fig. 4.10.

The main scenarios for the users of Restmarks include storing online one's bookmarks, sharing them with the Restmarks community and searching for bookmarks using tags. To apply our methodology, we created the association rules using the remaining 29 projects and isolated the rules that are activated by the annotated requirements of Restmarks, including their corresponding support and confidence values.

After that, we carefully examined all recommended requirements and annotated them as correct or incorrect, according to whether they are rational in the context of the project. The results indicate that most recommendations are quite rational. For example, the option to edit one's tags or to log out from one's account probably has been omitted by the engineers/stakeholders that originally compiled the requirements of Restmarks. Interestingly, the quality of each recommendation seems to be correlated with the corresponding support and confidence values. By visualizing the number of recommended requirements for each different combination of support and confidence values (in red color), including the percentage of correctly recommended requirements for these combinations (in blue color), we conclude that most recommendations exhibit high confidence. For example, we may note that the recommended requirements of Restmarks with confidence equal to 1 are always correct (upper part of Fig. 4.10c). The recommendations for requirements with lower confidence are in this case fewer (as is shown by the smaller circles in Fig. 4.10c); however, we may see that requirements with confidence values near 0.6/0.7 may also be correct recommendations, as long as they have relatively high support values.

Additionally, we employed a cross-validation scheme to further evaluate our approach and explore the influence of the support and confidence metrics on the quality of the recommendations. We split the dataset into six equal bins of five projects. For each bin, we removed the five projects of the bin from the dataset, we extracted the association rules from the remaining 25 projects and recommended new requirements for the 5 removed projects. After this procedure, we examined the recommended requirements for each project and determined whether each of them could be valid. The accumulated results of our evaluation for all projects are shown in Table 4.5 and visualized in Fig. 4.11.

In total, our system recommended 297 requirements, out of which we found 177 to be correct recommendations. Given that almost 60% of the recommendations can lead to useful requirements, we deem the results as satisfactory. Given a project, the requirements engineer would have been presented with a set of 10 requirements, out

A user must be able to create a user account by providing a username and a password.
A user must be able to login to his/her account by providing his username and password.
A user that is logged in to his/her account must be able to update his/her password.
A logged in user must be able to add a new bookmark to his/her account.
A logged in user must be able to retrieve any bookmark from his/her account.
A logged in user must be able to delete any bookmark from his/her account.
A logged in user must be able to update any bookmark from his/her account.
A logged in user must be able to mark his/her bookmarks as public or private.
A logged in user must be able to add tags to his/her bookmarks.
Any user must be able to retrieve the public bookmarks of any RESTMARKS's community user.
Any user must be able to search by tag the public bookmarks of a specific RESTMARKS's user.
Any user must be able to search by tag the public bookmarks of all RESTMARKS users.
A logged in user must be able to search by tag his/her private bookmarks as well.

(a)

+ The user must be able to edit bookmark. ($\sigma = 0.138, c = 1.0$)
+ The user must be able to view bookmark. ($\sigma = 0.103, c = 1.0$)
+ The user must be able to view account. ($\sigma = 0.103, c = 1.0$)
+ The user must be able to edit tag. ($\sigma = 0.103, c = 1.0$)
+ The user must be able to edit account. ($\sigma = 0.103, c = 1.0$)
+ The user must be able to logout account. ($\sigma = 0.172, c = 0.62$)
− The user must be able to contact account. ($\sigma = 0.172, c = 0.62$)
− The user must be able to contact bookmark. ($\sigma = 0.138, c = 0.5$)
− The user must be able to stop account. ($\sigma = 0.138, c = 0.5$)

$+/-$: Correctly/Incorrectly Recommended Requirement
σ: Support, c: Confidence

(b)

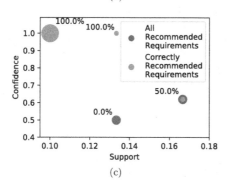

(c)

Fig. 4.10 Example depicting **a** the functional requirements of Restmarks, **b** the recommended requirements, and **c** the recommended requirements visualized

of which he/she would have selected 6 to add to the project. Most recommended requirements are extracted from rules with low support, which is actually expected since our dataset is largely domain-agnostic. However, low support rules do not necessarily result in low-quality recommendations, as long as their confidence is large enough. Indicatively, 2 out of 3 recommendations extracted from rules with confidence values equal to 0.5 may not be useful. However, setting the confidence

Table 4.5 Evaluation results for the recommended requirements

Support (σ)	Confidence (c)	# Correctly rec. requirements	Recommended requirements	Correctly rec. requirements (%)
0.2	1.0	1	2	50.0
0.133	1.0	23	37	62.16
0.1	1.0	43	76	56.58
0.133	0.8	2	4	50.0
0.2	0.75	0	1	0.0
0.1	0.75	64	92	69.57
0.167	0.71	0	1	0.0
0.133	0.67	13	14	92.86
0.167	0.62	1	1	100.0
0.1	0.6	9	17	52.94
0.133	0.57	4	7	57.14
0.167	0.56	4	5	80.0
0.1	0.5	13	40	32.5
Total		177	297	59.6

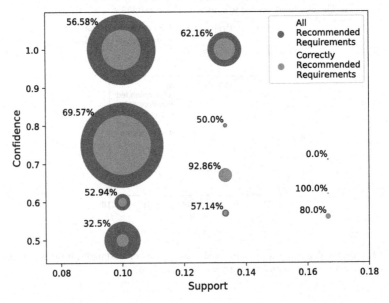

Fig. 4.11 Visualization of recommended requirements including the percentage of the correctly recommended requirements given support and confidence

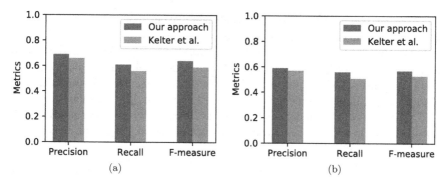

Fig. 4.12 Classification results of our approach and the approach of Kelter et al. for **a** use case diagrams and **b** activity diagrams

value to 0.75 ensures that more than 2 out of 3 recommended requirements will be added to the project.

4.5.2 UML Models Mining

To evaluate our UML mining model, we use a dataset of 65 use case diagrams and 72 activity diagrams, originating from software projects with different semantics. Our methodology involves finding similar diagrams and subsequently providing recommendations, and thus we initially split diagrams into six categories, including health/mobility, traffic/transportation, social/p2p networks, account/product services, business process, and generic diagrams. We construct all possible pairs of use case and activity diagrams, which are 2080 and 2556 pairs, respectively, and mark each pair as relevant or non-relevant according to their categories.

We compare our approach to that of Kelter et al. [26]. Given that the two approaches are structurally similar, their execution on use case diagrams provides an assessment of the semantic methodology. Concerning activity diagrams, the approach of Kelter et al. uses a static data model, and thus our evaluation should reveal how using a dynamic flow model can be more effective. Upon executing the approaches on the sets of use case and activity diagrams, we normalized the matching scores according to their average (so that pairs are defined when their score is higher than 0.5) and computed the precision, the recall, and the F-measure for each approach and for each diagram type. The results are shown in Fig. 4.12.

It is clear that our approach outperforms that of Kelter et al. when it comes down to finding relevant diagram pairs both for use case and for activity diagrams. The higher recall values indicate that our methodology can effectively retrieve more relevant diagram pairs, while high precision values indicate that the retrieved pairs are indeed relevant and false positives (non-relevant pairs that are considered relevant) are fewer. The combined F-measure also indicates that our approach is more effective for extracting semantically similar diagram pairs.

Examining the results indicates that our approach is much more effective when it comes down to identifying diagram pairs that are semantically similar. For example, out of the 2080 use case diagram pairs, 1278 were identified correctly as semantically similar (or not similar), while the approach of Kelter et al. [26] identified correctly 1167 pairs. The corresponding numbers for the 2556 activity diagram pairs are 1428 and 1310, for our approach and the approach of Kelter et al. [26], respectively.

4.6 Conclusion

Although several research efforts have focused on the area of requirements mining, most of these efforts are oriented toward high-level requirement representations and/or are dependent on domain-specific knowledge. In this chapter, we presented a mining methodology that can aid requirement engineers and stakeholders in the elicitation of functional requirements and the UML models for the system under development.

Concerning functional requirements, our system employs association rule mining to extract useful rules and uses these rules to create recommendations for the project under development. As we have shown in the evaluation section, our system can provide a set of recommendations out of which approximately 60% are rational. Thus, given that the requirements engineer (or a stakeholder) has compiled a set of requirements, he/she could effectively check whether he/she has omitted any important ones.

Our methodology for UML model mining employs domain-agnostic semantics and uses practical representations for UML use case and activity diagrams. As such, it covers both the static and the dynamic views of software projects, while the specifics of use case and activity diagrams, i.e., set-based and flow-based structure, are taken into account. The results of our evaluation indicate that our mining approach outperforms current state-of-the-art approaches.

Concerning future work, an interesting problem in the area of requirements elicitation is that of combining the preferences of different stakeholders. In the context of functional requirements, an extension of our system may be used in a multiple stakeholder scenario, where the different stakeholder requirements can be combined and a more project-specific set of association rules can be extracted. Another interesting future direction would be to improve the semantic matching module by incorporating information from online software databases (e.g., GitHub). Finally, our UML model mining methodology can include parsing different diagram types, such as UML class or sequence diagrams.

References

1. Montequin VR, Cousillas S, Ortega F, Villanueva J (2014) Analysis of the success factors and failure causes in Information & Communication Technology (ICT) projects in Spain. Procedia Technol 16:992–999
2. Leffingwell D (1997) Calculating your return on investment from more effective requirements management. Am Program 10(4):13–16
3. Thummalapenta S, Xie T (2007) PARSEWeb: a programmer assistant for reusing open source code on the web. In: Proceedings of the 22nd IEEE/ACM international conference on automated software engineering, ASE'07. ACM, New York, pp 204–213
4. Sahavechaphan N, Claypool K (2006) XSnippet: mining for sample code. SIGPLAN Not 41(10):413–430
5. Hummel O, Janjic W, Atkinson C (2008) Code conjurer: pulling reusable software out of thin air. IEEE Softw 25(5):45–52
6. Diamantopoulos T, Thomopoulos K, Symeonidis AL (2016) QualBoa: reusability-aware recommendations of source code components. In: Proceedings of the IEEE/ACM 13th working conference on mining software repositories, MSR'16, pp 488–491
7. Papamichail M, Diamantopoulos T, Symeonidis AL (2016) User-perceived source code quality estimation based on static analysis metrics. In: Proceedings of the 2016 IEEE international conference on software quality, reliability and security, QRS. Vienna, Austria, pp 100–107
8. Dimaridou V, Kyprianidis A-C, Papamichail M, Diamantopoulos T, Symeonidis A (2017) Towards modeling the user-perceived quality of source code using static analysis metrics. In: Proceedings of the 12th international conference on software technologies - volume 1, ICSOFT. INSTICC, SciTePress, Setubal, Portugal, pp 73–84
9. Luisa M, Mariangela F, Pierluigi NI (2004) Market research for requirements analysis using linguistic tools. Requir Eng 9(1):40–56, 2004
10. Kaindl H, Smialek M, Svetinovic D, Ambroziewicz A, Bojarski J, Nowakowski W, Straszak T, Schwarz H, Bildhauer D, Brogan JP, Mukasa KS, Wolter K, Krebs T (2007) Requirements specification language definition: defining the ReDSeeDS languages, deliverable D2.4.1. Public deliverable, ReDSeeDS (Requirements driven software development system) project
11. Smialek M (2012) Facilitating transition from requirements to code with the ReDSeeDS tool. In: Proceedings of the 2012 IEEE 20th international requirements engineering conference (RE), RE'12. IEEE Computer Society, Washington, pp 321–322
12. Wynne M, Hellesoy A (2012) The cucumber book: behaviour-driven development for testers and developers. Pragmatic Bookshelf, Raleigh
13. Mylopoulos J, Castro J, Kolp M (2000) Tropos: a framework for requirements-driven software development. Information systems engineering: state of the art and research themes. Springer, Berlin, pp 261–273
14. Boehm B, Basili VR (2001) Software defect reduction top 10 list. Computer 34:135–137
15. Kumar N, Ajmeri N, Ghaisas S (2010) Towards knowledge assisted agile requirements evolution. In: Proceedings of the 2nd international workshop on recommendation systems for software engineering, RSSE'10. ACM, New York, pp 16–20
16. Ghaisas S, Ajmeri N (2013) Knowledge-assisted ontology-based requirements evolution. In: Maalej W, Thurimella AK (eds) Managing requirements knowledge. Springer, Berlin, pp 143–167
17. Chen K, Zhang W, Zhao H, Mei H (2005) An approach to constructing feature models based on requirements clustering. In: Proceedings of the 13th IEEE international conference on requirements engineering, RE'05. IEEE Computer Society, Washington, pp 31–40
18. Alves V, Schwanninger C, Barbosa L, Rashid A, Sawyer P, Rayson P, Pohl C, Rummler A (2008) An exploratory study of information retrieval techniques in domain analysis. In: Proceedings of the 2008 12th international software product line conference, SPLC'08. IEEE Computer Society, Washington, pp 67–76
19. Frakes W, Prieto-Diaz R, Fox C (1998) DARE: domain analysis and reuse environment. Ann Softw Eng 5:125–141

20. Felfernig A, Schubert M, Mandl M, Ricci F, Maalej W (2010) Recommendation and decision technologies for requirements engineering. In: Proceedings of the 2nd international workshop on recommendation systems for software engineering, RSSE'10. ACM, New York, pp 11–15
21. Maalej W, Thurimella AK (2009) Towards a research agenda for recommendation systems in requirements engineering. In: Proceedings of the 2009 2nd international workshop on managing requirements knowledge, MARK'09. IEEE Computer Society, Washington, pp 32–39
22. Gomes P, Gandola P, Cordeiro J (2007) Helping software engineers reusing uml class diagrams. In: Proceedings of the 7th international conference on case-based reasoning: case-based reasoning research and development, ICCBR'07. Springer, Berlin, pp 449–462
23. Robles K, Fraga A, Morato J, Llorens J (2012) Towards an ontology-based retrieval of uml class diagrams. Inf Softw Technol 54(1):72–86
24. Alspaugh TA, Antón AI, Barnes T, Mott BW (1999) An integrated scenario management strategy. In: Proceedings of the 4th IEEE international symposium on requirements engineering, RE'99. IEEE Computer Society, Washington, pp 142–149
25. Blok MC, Cybulski JL (1998) Reusing uml specifications in a constrained application domain. In: Proceedings of the 5th Asia Pacific software engineering conference, APSEC'98. IEEE Computer Society, Washington, p 196
26. Kelter U, Wehren J, Niere J (2005) A generic difference algorithm for uml models. In: Liggesmeyer P, Pohl K, Goedicke M (eds) Software engineering, vol 64 of *LNI*. GI, pp 105–116
27. Diamantopoulos T, Symeonidis A (2017) Enhancing requirements reusability through semantic modeling and data mining techniques. Enterprise information systems, pp 1–22
28. Dumitru H, Gibiec M, Hariri N, Cleland-Huang J, Mo-basher B, Castro-Herrera C, Mirakhorli M (2011) On-demand feature recommendations derived from mining public product descriptions. In: Proceedings of the 33rd international conference on software engineering, ICSE'11. ACM, New York, pp 181–190
29. Lim SL, Finkelstein A (2012) StakeRare: using social networks and collaborative filtering for large-scale requirements elicitation. IEEE Trans Softw Eng 38(3):707–735
30. Castro-Herrera C, Duan C, Cleland-Huang J, Mobasher B (2008) Using data mining and recommender systems to facilitate large-scale, open, and inclusive requirements elicitation processes. In: Proceedings of the 2008 16th IEEE international requirements engineering conference, RE'08. IEEE Computer Society, Washington, pp 165–168
31. Mobasher B, Cleland-Huang J (2011) Recommender systems in requirements engineering. AI Mag 32(3):81–89
32. Konstan JA, Miller BN, Maltz D, Herlocker JL, Gordon LR, Riedl J (1997) GroupLens: applying collaborative filtering to usenet news. Commun ACM 40(3):77–87
33. Romero-Mariona J, Ziv H, Richardson DJ (2008) SRRS: a recommendation system for security requirements. In: Proceedings of the 2008 international workshop on recommendation systems for software engineering, RSSE'08. ACM, New York, pp 50–52
34. Woo HG, Robinson WN (2002) Reuse of scenario specifications using an automated relational learner: a lightweight approach. In: Proceedings of the 10th anniversary IEEE joint international conference on requirements engineering, RE'02. IEEE Computer Society, Washington, pp 173–180
35. Robinson WN, Woo HG (2004) Finding reusable uml sequence diagrams automatically. IEEE Softw 21(5):60–67
36. Bildhauer D, Horn T, Ebert J (2009) Similarity-driven software reuse. In: Proceedings of the 2009 ICSE workshop on comparison and versioning of software models, CVSM'09. IEEE Computer Society, Washington, pp 31–36
37. Salami HO, Ahmed M (2013) Class diagram retrieval using genetic algorithm. In: Proceedings of the 2013 12th international conference on machine learning and applications - volume 02, ICMLA'13. IEEE Computer Society, Washington, pp 96–101
38. Park W-J, Bae D-H (2011) A two-stage framework for uml specification matching. Inf Softw Technol 53(3):230–244
39. Chawathe SS, Rajaraman A, Garcia-Molina H, Widom J (1996) Change detection in hierarchically structured information. SIGMOD Rec 25(2):493–504

40. Wang Y, DeWitt DJ, Cai JY (2003) X-Diff: an effective change detection algorithm for XML documents. In: Proceedings 19th international conference on data engineering (Cat. No.03CH37405), pp 519–530
41. Bonilla-Morales B, Crespo S, Clunie C (2012) Reuse of use cases diagrams: an approach based on ontologies and semantic web technologies. Int J Comput Sci 9(1):24–29
42. Miller GA (1995) WordNet: a lexical database for english. Commun ACM 38(11):39–41
43. Finlayson M (2014) Java libraries for accessing the princeton wordnet: comparison and evaluation. In: Orav H, Fellbaum C, Vossen P (eds) Proceedings of the 7th global wordnet conference. Tartu, Estonia, pp 78–85
44. Pedersen T, Patwardhan S, Michelizzi J (2004) WordNet:: similarity: measuring the relatedness of concepts. In: Demonstration papers at HLT-NAACL 2004, HLT-NAACL–Demonstrations'04. Association for Computational Linguistics, Stroudsburg, pp 38–41
45. Lin D (1998) An information-theoretic definition of similarity. In: Proceedings of the 15th international conference on machine learning, ICML'98. Morgan Kaufmann Publishers Inc, San Francisco, pp 296–304
46. Agrawal R, Imieliński T, Swami A (1993) Mining association rules between sets of items in large databases. In: Proceedings of the 1993 ACM SIGMOD international conference on management of data, SIGMOD'93. ACM, New York, pp 207–216
47. Agrawal R, Srikant R (1994) Fast algorithms for mining association rules in large databases. In: Proceedings of the 20th international conference on very large data bases, VLDB'94. Morgan Kaufmann Publishers Inc, San Francisco, pp 487–499
48. Cormen TH, Leiserson CE, Rivest RL, Stein C (2009) Introduction to algorithms, 3rd edn. The MIT Press, Cambridge, pp 390–396

Part III
Source Code Mining

Chapter 5
Source Code Indexing for Component Reuse

5.1 Overview

The introduction of the open-source development paradigm has changed the way software is developed and distributed. Several researchers and practitioners now develop software as open source and store their projects online. This trend has acquired momentum during the last decade, leading to the development of numerous online software repositories, such as GitHub[1] and Sourceforge.[2] One of the main arguments in favor of open-source software development is that it can be improved by harnessing the available software repositories for useful code and development practices.

Another strong argument is that software developers can greatly benefit from software reuse, in order to reduce the time spent for development, as well as to improve the quality of their projects. Nowadays, the development process involves searching for reusable code in general-purpose search engines (e.g., Google or Yahoo), in open-source software repositories (e.g., GitHub), or even in question-answering communities (e.g., Stack Overflow[3]). Indeed, the need for searching and reusing software components has led to the design of search engines specialized in code search. *Code Search Engines (CSEs)* improve search in source code repositories, using keywords, or even using syntax-aware queries (e.g., searching for a method with specific parameters). Another category of more specialized systems is that of *Recommendation Systems in Software Engineering (RSSEs)*. In the context of code reuse, RSSEs are intelligent systems, capable of extracting the query from the source code of the developer and processing the results of CSEs to recommend software components, accompanied by information that would help the developer understand and reuse the code. In this chapter, we focus on CSEs, while in the next we focus on RSSEs.

---

[1] https://github.com/.

[2] https://sourceforge.net/.

[3] http://stackoverflow.com/.

© Springer Nature Switzerland AG 2020

T. Diamantopoulos and A. L. Symeonidis, *Mining Software Engineering Data for Software Reuse*, Advanced Information and Knowledge Processing, https://doi.org/10.1007/978-3-030-30106-4_5

Although contemporary CSEs are certainly gaining their share in the developer market, the proper design of a system that addresses the needs of developers directly is still an open research question. In specific, current CSEs fall short in one or more of the following three important aspects: (a) they do not offer flexible options to allow performing advanced queries (e.g., syntax-aware queries, regular expressions, snippet, or project search), (b) they are not updated frequently enough to address the ad hoc needs of developers, since they are not integrated with hosting sites, and (c) they do not offer complete well-documented APIs, to allow their integration with other tools. The latter offering is very important also for RSSEs; their aim to offer recommendations to developers in a more personalized context is strongly tied to their need for using specialized open-API CSEs in the backend and has been highlighted by several researchers [1–4].

In this chapter, we present AGORA,[4] a CSE that satisfies the aforementioned requirements [5]. AGORA indexes all source code elements, while the code is also indexed at token level and the hierarchy of projects and files is preserved. Thus, it enables syntax-aware search, i.e., searching for classes with specific methods, etc., and allows searching using regular expressions, searching for snippets or for multiple objects (at project level). The stored projects are continuously updated owing to the integration of AGORA with GitHub. Finally, AGORA offers a REST API that allows integration with other tools.

5.2   Background on Code Search Engines

5.2.1   Code Search Engines Criteria

Given the status quo in software engineering practices and the research on search engines for source code (see next subsection), CSEs should abide by the following criteria:

- **Syntax-awareness**: The system must support queries given specific elements of a file, e.g., class name, method name, return type of method, etc.
- **Regular expressions**: Regular expressions have to be supported so that the user can search for a particular pattern.
- **Snippet search**: Although most CSEs support free text search, this is usually ineffective for multi-line queries. A CSE should allow searching for small blocks of code, known as snippets.
- **Project search**: The system must allow the user to search for projects given a file query. In other words, the system must store the data with some kind of structure

---

[4]The term "agora" refers to the central spot of ancient Greek city-states. Although it is roughly translated to the English term "market", "agora" can be better viewed as an assembly, a place where people met not only to trade goods, but also to exchange ideas. It is a good fit for our search engine since we envision it as a place where developers can (freely) distribute and exchange their source code, and subsequently their ideas.

(schema) so that complex queries of the type "find project that has file" can be answered.

- **Integration with hosting websites**: Importing projects into the CSE repository and keeping the index up-to-date should be straightforward. Hence, CSEs must integrate well with source code hosting websites (e.g., GitHub).
- **API**: The system must have a public API. Although this criterion may seem redundant, it is crucial because a comprehensive public API allows different ways of using the system, e.g., by creating a UI, by some IDE plugin, etc.

Note that the list of desired features/criteria outlined above covers the most common aspects of CSEs; there are several other features that could be considered in a code search scenario, such as search for documentation. However, we argue that the selected list of criteria outlines a reasonable set of features for code search scenarios and especially for integrating with RSSEs, a view also supported by current research [3, 6]. In the following subsections, we provide a historical review for the area of CSEs and discuss the shortcomings of well-known CSEs with respect to the above criteria. This way we justify our rationale for creating a new CSE.

5.2.2 Historical Review of Code Search Engines

Given the vastness of code in online repositories (as already analyzed in Chap. 2), the main problem is how to search for and successfully retrieve code according to one's needs. Some of the most popular CSEs are shown in Table 5.1.

An important question explored in this chapter is whether these CSEs are actually as effective and useful as they can be for the developer. Codase, one of the first engines with support for syntax-aware search (e.g., search for class or method name), seemed promising; however, it didn't go past the beta phase. Krugle Open Search, on the other hand, is one of the first CSEs that is still active. The main advantage of Krugle is its high-quality results; the service offers multiple search options (e.g., syntax-aware queries) on a small but well-chosen number of projects, including Apache Ant, Jython, etc. Although Krugle does indeed cover criteria 1 and 3, it does not integrate well with hosting sites, and it does not offer a public API.

In 2006, one of the big players, Google, decided to enter the CSE arena. The Google Code Search Engine was fast, supported regular expressions, and had over 250000 projects to search for source code. Its few shortcomings were spotted in criteria 3 and 4 since the engine did not allow snippets or complex intra-project queries. However, in 2011 Google announced its forthcoming shutdown[5] to finally shut it down in 2013. Despite its shutdown, the service has increased the interest for CSEs, leading developers to search for alternatives.

One such early alternative was MeroBase [6]. The index of MeroBase contained projects as software artifacts and supported searching and retrieving source code and

---

[5] Source: http://googleblog.blogspot.gr/2011/10/fall-sweep.html.

Table 5.1 Popular source code search engines

Search engine	Year	Website	# Projects
Codase	2005–2014	http://www.codase.com/	>10K
Krugle	2006–today	http://opensearch.krugle.org/	~500
Google CSE	2006–2013	http://www.google.com/codesearch	>250K
MeroBase	2007–2014	http://www.merobase.com/	>5K
Sourcerer	2007–2012	http://sourcerer.ics.uci.edu/	>70K
GitHub CSE	2008–today	https://github.com/search	~14.2M
BlackDuck	2008–2016	https://code.openhub.net/	>650K
Searchcode	2010–today	https://searchcode.com/	>300K

This table includes only CSEs, i.e., search engines that allow searching for source code. Thus, engines such as GrepCode (http://grepcode.com/), NerdyData (http://nerdydata.com/), or SymbolHound (http://symbolhound.com/) are not included, since their functionality is to search for projects (in the case of GrepCode), website-only code (in the case of NerdyData), or for answers in question-answering sites (in the case of SymbolHound)

projects according to several criteria, including source code criteria (e.g., class and method names), licensing criteria, the author of the artifact, etc. As noted by the authors [6], the main weakness of MeroBase is its flat structure, which is hard to change given the underlying indexing. Due to this, MeroBase does not support the advanced search capabilities described in criteria 2 and 3, while its index, though large, is not updatable using a hosting site. A similar proposal is Sourcerer [7], which employs a schema with multiple syntax-aware features extracted from source code and thus supports different types of queries. However, Sourcerer is also not integrated with code repositories, and thus its index has to be manually updated.

GitHub, the most popular source code hosting site today, has its own search engine for its open-source repositories. GitHub Search is fast and reliable, providing up-to-date results not only for source code, but also for repository statistics (commits, forks, etc.). Its greatest strength when it comes to source code is its huge project index, while its API is well defined. However, its search options are rather outdated; there is no support for regular expressions or generally syntax-aware queries. As a result, it covers none of the first four criteria.

BlackDuck Open Hub is also one of the most popular CSEs with a quite large index. BlackDuck acquired the former CSE of Koders in 2008 and merged it with the index of Ohloh (obtained in 2010) to finally provide Ohloh Code in 2012. Although

Table 5.2 Feature comparison of popular source code search engines and AGORA

Search engine	Criterion 1 syntax-awareness	Criterion 2 regular expressions	Criterion 3 snippet search	Criterion 4 project search	Criterion 5 repositories integration	Criterion 6 API support
AGORA	✓	✓	✓	✓	✓	✓
Codase	✓	×	×	×	×	✓
Krugle	✓	×	✓	×	×	×
Google CSE	✓	✓	×	×	✓	✓
MeroBase	✓	×	×	✓	×	✓
Sourcerer	✓	✓	×	✓	×	✓
GitHub CSE	×	×	×	×	✓	✓
BlackDuck	✓	✓	×	×	×	×
Searchcode	×	✓	×	×	✓	✓

✓ Feature supported; ×Feature not supported; ✓ Feature partially supported

the service initially aimed at filling the gap left by Google Code Search,[6] it later grew to become a comprehensive directory of source code projects, providing statistics, license information, etc. Ohloh Code was renamed to BlackDuck Open Hub Code Search in 2014. The service has support for syntax-aware queries and offers plugins for known IDEs (Eclipse, Microsoft Visual Studio). However, it does not support snippet and project search (criteria 3 and 4).

Finally, a promising newcomer in the field of CSEs is Searchcode. Under its original name (search[co.de]) the service offered a documentation search for popular libraries. Today, it crawls several known repositories (GitHub, Sourceforge, etc.) for source code, and thus its integration with source code hosting sites is quite advantageous. However, the engine does not support syntax-aware queries, snippet, or project search, and hence criteria 1, 3, and 4 are not met. Conclusively, none of the CSEs presented in Table 5.1 (and no other to the best of our knowledge) covers adequately all of the above criteria.

5.2.3 AGORA Offerings

The effectiveness of a CSE depends on several factors, such as the quality of the indexed projects and the way these projects are indexed. We compare AGORA with other known CSEs and draw useful conclusions. In particular, since the scope of this chapter lies in creating a CSE that meets the criteria defined in Sect. 5.2.1, we evaluate whether AGORA fulfills these criteria compared to other state-of-the-art engines. This comparison is shown in Table 5.2.

---

[6]Source: http://techcrunch.com/2012/07/20/ohloh-wants-to-fill-the-gap-left-by-google-code-search/.

Concerning the search-related features, i.e., criteria 1–4, AGORA is the CSE that allows the most flexible queries. It supports syntax-aware queries, regular expressions, even searching for snippets and projects. Other CSEs, such as Codase or Black-Duck, cover the syntax-awareness and regular expression criteria, without, however, support for snippet or project search. Krugle supports searching for snippets but does not allow using regular expressions or searching for projects. MeroBase and Sourcerer are the only other CSEs (except from AGORA) to partially support complex project search, aiming, however, mostly at the documentation of the projects.

Integration with hosting websites is another neglected feature of modern CSEs. GitHub Search or the discontinued Google Code Search obviously supports it since the hosting sites are actually in their own servers (GitHub and Google Code, respectively). Other than these services, though, Searchcode is the only other service to offer an index that is integrated with hosting sites, namely, Github, Sourceforge, etc.

The sixth criterion is also quite important since CSEs with comprehensive APIs can easily be extended by building web UIs or IDE plugins. Most CSEs fulfill this criterion, with GitHub and Searchcode having two of the most comprehensive APIs. The API of AGORA, which actually wraps the API provided by Elasticsearch, is quite robust and allows performing different types of queries.

We do not claim to have created an overall stronger CSE; this would also require evaluating several aspects, such as the size of its index or how many languages it supports; AGORA is still experimental. However, its architecture for both the language and the web hosting facilities is mostly agnostic; adding new languages and/or integrating with new code hosting services (other than GitHub) is feasible. Practically, AGORA is a solution suitable for supporting research in software reuse.

5.3 The AGORA Code Search Engine

5.3.1 Overview

The overall architecture of AGORA is shown in Fig. 5.1. ElasticSearch [8] was selected as our search server since it abstracts Lucene well, and allows harnessing its power, while at the same time overcoming any difficulties coming from Lucene's "flat" structure. Storage in ElasticSearch is quite similar to the NoSQL document-based paradigm. Objects are stored in documents (the equivalent of records in a relational schema), documents are stored inside *collections* (the equivalent of relational tables), and collections are stored in *indexes* (the equivalent of relational databases). We define two collections: *projects* and *files*. Projects contain information about software projects, while files contain information about source code files.

As depicted in the figure, the *Elasticsearch Index* is accessible through the *Elasticsearch Server*. In addition, an *Apache Server* is instantiated in order to control all communication to the server. The Apache server allows only authorized input to the index, thus ensuring that only the administrator can perform specific actions

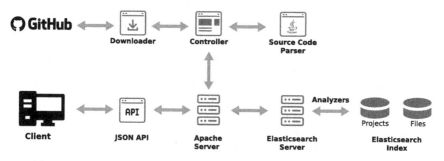

Fig. 5.1 The architecture of AGORA

(e.g., delete a document, backup the index, etc.). For the end users of AGORA, the server is accessible using the comprehensible *JSON API* offered by Elasticsearch.

The *controller* shown in Fig. 5.1 is one of the most important components of AGORA. Built in Python, the controller provides an interface for adding GitHub projects in ElasticSearch. In specific, given a GitHub project address, the controller provides the address to the *downloader*. The downloader checks whether the source code of the project is already downloaded, and downloads or updates it. Typically, the downloader comprises two components, the *GitHub API Downloader* and the *Git Downloader*. The GitHub API Downloader fetches information about a GitHub project using the GitHub API [9], including the git address of the project and its source code tree. The Git Downloader is actually a wrapper for the Git tool; it clones GitHub projects or pulls them if they already exist. Given the tree provided by the API, the downloader can provide the controller with a list of files and sha ids. After that, the sha ids are used by the controller in order to determine which of the files need to be added, updated, or deleted from the index.

Another important step is performed before adding the new documents to the index. The source code of the project is analyzed using the *Source Code Parser*. The Source Code Parser receives as input a source code file and extracts its *signature*. The signature of a file is represented as a JSON document containing all elements defined or used in the file, e.g., class names, method names, etc.

The JSON documents are added to the projects and the files index. The schemata of the fields are dictated by the *mappings*. AGORA has two mappings, one for the JSON documents of the projects index and one for those of the files index. When a document is added to the index, its fields (JSON entries) are analyzed using an *analyzer*. This step is important since it allows Elasticsearch to perform fast queries over the analyzed fields.

The index of AGORA can obviously be used to store projects written in different programming languages. In specific, adding source code files of a project written in a specific programming language involves three tasks: (a) developing a Source Code Parser (to extract information from the source code), (b) creating appropriate analyzers that conform to the specifics of the "new" programming language, and (c) designing a structure that shall store the elements of the source code included in each file. Without loss of generality, in the following paragraphs, we illustrate how these tasks can be performed for the Java programming language.

The Source Code Parser for Java projects is built using the Java Compiler Tree API [10] offered by Oracle. It receives as input one or more files and traverses the *Abstract Syntax Tree (AST)* of each file in order to output all the Java-specific elements defined or used in the file: packages, imports, classes, methods, variables, etc. Additionally, it contains information about these elements, such as the scope of variables and the return type of methods. Concerning the analyzers of Elasticsearch, these also have to conform to the characteristics of Java files (e.g., support indexing camel case terms). The analyzers that we have developed for the Java language are presented in Sect. 5.3.2, while Sect. 5.3.3 illustrates the mappings used for storing projects and files.

5.3.2 Analyzers

Field analysis is one of the most important features of indexing servers, since it affects the quality and efficiency of the returned results. In Elasticsearch, analyzers comprise a single *tokenizer* and zero or more *token filters*. Given a field, the tokenizer splits its value into tokens and the token filters accept a stream of tokens which they can modify, delete, or even add. For example, in a text search engine scenario, a tokenizer would split a document into words according to spaces and punctuation, and then it would probably apply a token filter that makes all words lowercase and another one that would remove stop words. Then all fields would be stored in the index in tokens, thus allowing fast and accurate queries. The same approach holds for source code analysis.

Elasticsearch has several default analyzers and even allows creating custom ones given its collection of tokenizers and token filters. For our scenario, we used the *standard* analyzer offered by Elasticsearch and created also three custom analyzers: the *Filepath* analyzer, which is a custom analyzer for file paths, the *CamelCase* analyzer, which handles camel case tokens, and the *Javafile* analyzer, for analyzing java files.

5.3.2.1 Standard Analyzer

The standard analyzer initially employs the standard tokenizer of Elasticsearch in order to split text into tokens according to the Unicode Standard Annex #29 [11], i.e., on spaces, punctuation, special characters, etc. After that, the tokens are converted to lowercase using a Lowercase Token Filter and common words are removed using

```
#Match
([^\\p{L}\\d]+)                              #non-letters and numbers,
 | (?<=\\D)(?=\\d)                           #or non-number then number,
 | (?<=\\d)(?=\\D)                           #or number then non-number,
 | (?<=[ \\p{L} && [^\\p{Lu}]])               #or lower case
#followed by
(?=\\p{Lu})                                 #upper case,
 | (?<=\\p{Lu})                              #or upper case
   (?=\\p{Lu}[\\p{L}&&[^\\p{Lu}]])   #then upper and lower case
```

Fig. 5.2   The regular expression of the CamelCase analyzer

a Stop Token Filter.[7] The standard analyzer is the default analyzer of Elasticsearch, since it is effective for most fields.

5.3.2.2   Filepath Analyzer

The filepath analyzer is used to analyze file or Internet paths. It comprises the Path Hierarchy Tokenizer and the Lowercase Token Filter of Elasticsearch. The analyzer receives input in the form "/path/to/file" and outputs tokens "/path", "/path/to", and "/path/to/file".

5.3.2.3   CamelCase Analyzer

Although the standard analyzer is effective for most fields, the fields derived from source code files usually conform to the standards of the language. Since our targeted language is Java, we created an analyzer that parses camel case text. This is accomplished using a *pattern* tokenizer, which uses a regular expression pattern in order to split the text into tokens. For camel case text, we used the tokenizer shown in [12]. The pattern is shown in Fig. 5.2.

Upon splitting the text into tokens, our CamelCase analyzer transforms the tokens to lowercase using the Lowercase Token Filter of Elasticsearch. No stop words are removed.

5.3.2.4   Javafile Analyzer

The Javafile analyzer is used to analyze Java source code files. The analyzer employs the standard tokenizer and the Stop Token Filter of Elasticsearch. It splits the text

---

[7]Elasticsearch supports removing stop words for several languages. In our case, the default choice of English is adequate.

---

void, if, else, do, for, switch, case, break, while, default, true, false, return,
this, null, instanceof, try, catch, throws, throw, finally, super, assert, const,
package, import, implements, extends, synchronized, continue, interface,
class, public, private, protected, final, static, abstract, native, transient

---

Fig. 5.3 List of Java stop words

into tokens and the most common Java source file tokens are removed. The Java stop words used are shown in Fig. 5.3.

The stop word list of Fig. 5.3 contains all the terms that one would want to omit when searching for snippets. Note, however, that we decided to keep the types (e.g., "int", "float", "double") since they are useful for finding similar code snippets. The type "void", however, is removed since it is so frequent that it would practically add nothing to the final result.

5.3.3 Document Indexing

As noted in Sect. 5.3.1, Elasticsearch stores data in documents. The way each document is stored (i.e., its schema) and the way each field is analyzed by one of the analyzers of Sect. 5.3.2 are determined by its *mapping*. For AGORA and for the Java language, we defined two mappings, one for project documents and one for file documents. Additionally, we connected these two mappings to each other in a *parent-child* relationship, where each file document maps to one parent project document. We present the two mappings in the following paragraphs.

5.3.3.1 Indexing Projects

Projects are identified uniquely in AGORA by their name and username joined with a slash (/). Thus, the _id_ field of the mapping has values of the form "user/repo". The fields for project documents are shown in Table 5.3, along with an example instantiation of the mapping for the egit Eclipse plugin.

Most of the project's fields are useful to be stored, yet analyzing all of them does not provide any added value. Hence, URLs (the "url", the GitHub API "git_url") and "default_branch" are not analyzed. The field "fullname", however, which is the same as the "_id", is analyzed using the standard analyzer so that it is searchable. In addition, one can search by the "user" and "name" fields, corresponding to the owner and the project name, which are also analyzed using the standard analyzer.

Table 5.3 The mapping of projects documents of the AGORA index

Name	Type	Analyzer	Example
fullname	string	–	Eclipse/egit-github
branch	string	–	Master
url	string	–	https://api.github.com/repos/eclipse/egit-github
git_url	string	–	git://github.com/eclipse/egit-github.git
user	string	Standard	Eclipse
name	string	Standard	egit-github

Table 5.4 The mapping of files documents of the AGORA index

Name	Type	Analyzer	Example
fullpathname	string	Filepath	eclipse/egit-github/ .../DateFormatter.java
path	string	Filepath	org.../DateFormatter.java
name	string	Standard	DateFormatter.java
project	string	Standard	eclipse/egit-github
mode	string	–	100644
sha	string	–	e51f1efca2155fc7af3...
type	string	–	blob
extension	string	–	Java
url	string	–	https://api.github.com/ repos/eclipse/...
content	string	Standard	...
analyzedcontent code	string	Javafile	...
package	string	CamelCase	org.eclipse....client
imports	string	CamelCase	[java.util.TimeZone, java.util.Date, ...]
class	See Table 5.5		See Table 5.5
otherclasses	See Table 5.5		[]

5.3.3.2 Indexing Files

As already mentioned, we have structured the project and file mappings using the parent-child relation provided by Elasticsearch. Each file has a "_parent" field, which is a link to the respective document of the project. This field can be used to perform *has_child* queries (see Sect. 5.4.2.3) or retrieve all the files of a project. The "_id" of each file has to be unique; it is formed by the full path of the file including also the project, i.e, user/project/path/to/file. Table 5.4 contains the rest of the fields as well as an example for the DateFormatter.java file of the Eclipse egit plugin of Table 5.3.

As in projects, the "_id" is also copied to the analyzed field "fullpathname" so that it is searchable. Both this and the "path" field, which refers to the path without the project prefix, are analyzed using the filepath analyzer. The filename of each file ("name") and the name of the project it belongs to ("project") are analyzed using the standard analyzer.[8]

Again, some fields do not need to be searchable, yet are useful to be stored. The fields "mode" and "type", which refer to the git file mode[9] and the git type,[10] as well as the "extension" and the "url" of each file are not analyzed. The field "sha" is also not analyzed. However, it should be mentioned that "sha" is very important since it maintains an id for the content of the file. This field is used to check whether a file needs to be updated when updating a project.

Furthermore, since files also contain source code, several important fields are defined to cover the source code search. At first, the content of each file is analyzed with two analyzers, the standard analyzer and the Javafile analyzer, storing the results in the fields "content" and "analyzedcontent", respectively. This analysis allows performing strict text matching on the "content" field as well as loose search on "analyzedcontent" given the stop words on the second one are removed.

After that, the instances for the subfields of the field "code" are extracted using the Source Code Parser (see Fig. 5.1). Two such fields that are present in Java files are "package" and "imports" (which are stored as a list since they can be more than one). The two fields are analyzed using the CamelCase analyzer, since both the package of a Java file and its imports follow the Java coding conventions. Since each Java file has one public class and possibly other non-public classes, the mapping also contains the fields "class" and "otherclasses".

Note that "otherclasses" is a list of type "nested". This is a special type of field offered by Elasticsearch that allows querying subfields in a coherent way. In specific, given that a class has two subfields, "name" and "type" (class or interface), if we want to search for a specific name and type both for the same class, then the list must be "nested". Otherwise, our search may return a document that has two different classes, one having only the name and one having only the type that we searched for (more about nested search in Sect. 5.4.2.1). The subfields of a class are shown in Table 5.5, including an example for the DateFormatter class corresponding to the DateFormatter.java file of Table 5.4.

As depicted in Table 5.5, the "extends" and "implements" fields are analyzed using the CamelCase analyzer. Note that "implements" is of type "array", since in Java a class may implement more than one interface. A Java class can also have many methods, with each of them having its own name, parameters, etc. So the "methods" field is of type "nested" so that methods can be searched using an autonomous query, while also allowing multi-method queries. The "name" and the "returntype" of a method are string fields analyzed using the CamelCase analyzer, while the

---

[8]Note that the standard analyzer initially splits the filename into two parts, the filename without extension and the extension since it splits according to punctuation.

[9]One of 040000, 100644, 100664, 100755, 120000, or 160000 which correspond to directory, regular non-executable file, regular non-executable group-writeable file, regular executable file, symbolic link, or gitlink, respectively.

[10]One of blob, tree, commit, or tag.

Table 5.5 The inner mapping of a class for the files documents of AGORA

Name	Type	Analyzer	Example
extends	string	CamelCase	–
implements	array	CamelCase	[JsonSerializer<Date>, ...]
methods	nested		
modifiers	array	–	[public]
name	string	CamelCase	serialize
parameters	nested		
name	string	CamelCase	Date
type	string	CamelCase	Date
returntype	string	CamelCase	JsonElement
throws	array	CamelCase	[]
modifiers	array	–	[public]
name	string	CamelCase	DateFormatter
type	string	CamelCase	Class
variables	nested		
modifiers	array	–	[private, final]
name	string	CamelCase	Formats
type	string	CamelCase	DateFormat[]
innerclasses	(1-level) recursive of the class mapping		

exceptions ("throws") are stored in arrays and also indexed using the CamelCase analyzer. The field "modifiers" is also an array; however, it is not analyzed since its values are specific (one of public, private, final, etc.). The subfield "parameters" holds the parameters of the method, including the "name" and the "type" of each parameter, both analyzed using the CamelCase analyzer.[11]

The class "modifiers" are not analyzed, whereas their "name" is analyzed. The field "type" is also not analyzed since its values are specific (one of class, interface, type). The field "variables" holds the variables of classes in a "nested" type, and the subfields corresponding to the "name" and "type" of each variable are analyzed using the CamelCase analyzer. Finally, inner classes have the same mapping as in Table 5.5, however without containing an "innerclasses" field, thus allowing only one level of inner classes. Allowing more levels would be ineffective.

---

[11] The CamelCase analyzer is quite effective for fields including Java types; the types are conventionally in camelCase, while the primitives are not affected by the CamelCase tokenizer, i.e., the text "float" results in the token "float".

5.3.4 Relevance Scoring

The scoring scheme of Elasticsearch [13] is used to rank the retrieved results for a query according to their relevance. When forming a query, one may select any of the fields defined in the mappings of AGORA (see Sect. 5.3.3) or combinations among them. As already mentioned, all analyzed fields (using any one of the four defined analyzers) are indexed as split tokens, disregarding punctuation and stop words.

At first, the text of the query is also analyzed using the analyzer that corresponds to the addressed field. After that, the similarity between the terms/tokens of the query and the terms/tokens of the document field is computed using the *vector space model* [14]. In the vector space model, any term is a dimension, thus each document (or the query) can be represented as a vector. Given a collection of documents D, the relevance weight of a term t in a document d is computed using the *term frequency-inverse document frequency (tf-idf)*:

$$w_{t,d} = \sqrt{f(t,d)} \cdot \left(1 + log \frac{|D|}{|d \in D : t \in d| + 1}\right) \cdot \frac{1}{\sqrt{|d|}} \qquad (5.1)$$

The first term of the above product is the frequency of the term in the document (tf), computed as the square root of the number of times the term appears in the document $f(t,d)$. The second term is the inverse document frequency (idf), computed as the logarithm of the total number of documents $|D|$ divided by the number of documents that contain the term $|d \in D : t \in d|$. Finally, the third term is a field-length normalizer that accounts for the number of terms in the document $|d|$, and effectively penalizes larger documents. In the case of non-analyzed fields, the content of the field has to be exactly specified and the system will return a binary match-no match decision, i.e., the result of Eq. (5.1) will be either 1 or 0. Array fields are treated like already split tokens.

Thus, the document vectors are represented using the term weights defined by the above equation. The final score between a query vector q and a document vector d is computed using the cosine similarity between them:

$$score(q,d) = \frac{q \cdot d}{|q| \cdot |d|} = \frac{\sum_1^n w_{t_i,q} \cdot w_{t_i,d}}{\sum_1^n w_{t_i,q}^2 \cdot \sum_1^n w_{t_i,d}^2} \qquad (5.2)$$

where t_i is the ith term of the query and n is the total number of terms in the query.

Finally, when a query addresses more than one fields, an "and" relationship among the fields is implied. Then, assuming that the score of equation (5.2) is defined per field, i.e., $score(q, f)$ is the score for query q and field f, the score for each document d is computed using the scoring function:

$$score(q,d) = norm(q) \cdot coord(q,d) \cdot \sum_{t \in q} score(q,f) \qquad (5.3)$$

where $norm(q)$ is the query normalization factor, defined as the inverse square root of the sum of squared *idf* of each term in the query, i.e., $1/\sqrt{\sum_1^n w_{t_i,q}}$ and $coord(q, d)$ is the coordination factor defined as the number of matching query terms divided by the total number of terms in the query, i.e., $|t \in q : t \in d|/|q|$.

5.4 Searching for Code

5.4.1 AGORA Search Features

AGORA has a comprehensive REST API as well as a web UI, shown in Fig. 5.4. Although the API of AGORA allows performing any complex request supported by Elasticsearch, in practice we may distinguish among four types of searches, which are also the ones supported by the web interface: *Simple Search*, *Advanced Search*, *Snippet Search*, and *Project Search*.

The *Simple Search* feature involves the Elasticsearch field "_all", which allows searching for all the fields of a file (e.g., filename, content, etc.). This type of search further supports the use of regular expressions.

The *Advanced Search* functionality allows searching for a class with specific attributes including imports, variable names and types, method names, return types and their parameters' names and types, etc. Using this functionality, developers can easily find reusable components with specific inputs/outputs to integrate to their source code or even find out how to write code in context.

The *Snippet Search* allows providing a source code snippet as input in order to find similar code. Although using syntax-aware queries yields satisfactory results at class or method level, when it comes to finding API usage examples, searching for snippets can be quite helpful, as most API usage scenarios involve method calls within a few lines of code.

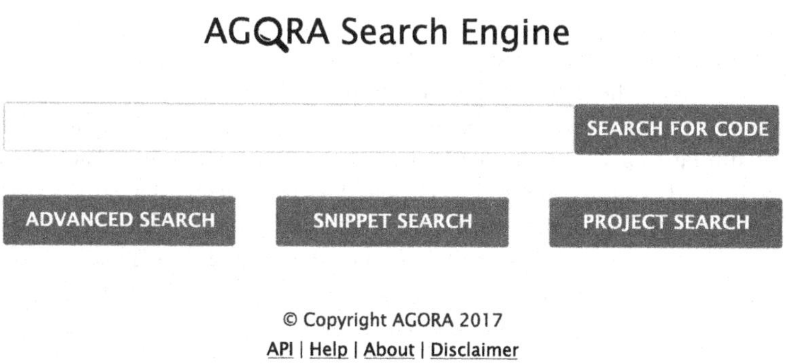

Fig. 5.4 Home page of AGORA

Finally, the *Project Search* functionality allows searching for software projects that involve certain components or follow a specific structure. Thus, one may search for projects that conform to specific design patterns [15] or architectures (e.g., MVC).

Currently, the index is populated given a list of the 3000 most *popular* GitHub projects, as determined by the number of stars, which, upon removing forked projects and invalid projects (e.g., without java code), resulted in 2709 projects. Preliminary research has indicated that highly rated projects (i.e., with large number of stars/forks) exhibit also high quality [16], have sufficient documentation/readme files [17, 18], and involve frequent maintenance releases, rewrites and releases related to both functional and non-functional requirements [19]. Upon further measuring the reusability of software components with static analysis metrics [20, 21], we could argue that these projects include reusable code; indicatively, we have shown that roughly 90% of the source code of the 100 most popular GitHub repositories is reusable and of high quality [20]. Given also that the projects are highly diverse, spanning from known large-scale projects (e.g., the Spring framework or the Clojure programming language) to small-scale projects (e.g., android apps), we may conclude that our index contains a variety of reusable solutions.

In Sect. 5.4.2, we provide example queries performed on the index of AGORA, indicating the corresponding usage scenarios and illustrating the syntax required to search in the AGORA API. Furthermore, in Sect. 5.4.3, we provide an example end-to-end scenario for using AGORA, this time through its web interface, to construct a project with three connected components.

5.4.2 AGORA Search Scenarios

5.4.2.1 Searching for a Reusable Component

One of the most common scenarios studied by current literature [3, 4, 6, 22] is that of searching for a reusable software component. Components can be seen as individual entities with specific inputs/outputs that can be reused. The simplest form of a Java component is a class. Thus, a simple scenario includes finding a class with specific functionality. Using the API of AGORA, finding a class with specific methods and/or variables is straightforward. For example, assume a developer wants to use a data structure similar to a stack with the known push/pop functionality. Then the query would be formed as shown in Fig. 5.5. The query is a *bool* query that allows searching using multiple conditions that must hold at the same time. In this case, the query involves three conditions: one for the name of the class which must be "stack", a nested one for a method with name "push" and return type "void", and another nested one for a method with name "pop" and return type "int". The response for the query is shown in Fig. 5.6.

On our installed snapshot of AGORA, the query took 109 ms and 412275 documents ("hits") were found. For each document, the "_score" is determined according to the scoring scheme defined in Sect. 5.3.4. In this case, the developer can examine

```
{
    "query": {"bool": {"should": [
        {"match": {"code.class.name": "stack"}},
        {"nested": {
            "path": "code.class.methods",
            "query": {"bool": {"should": [
                {"match": {"code.class.methods.name": "push"}},
                {"term": {"code.class.methods.returntype": "void"}}
            ]}}
        }},
        {"nested": {
            "path": "code.class.methods",
            "query": {"bool": {"should": [
                {"match": {"code.class.methods.name": "pop"}},
                {"term": {"code.class.methods.returntype": "int"}}
            ]}}
        }},
    ]}}
}
```

Fig. 5.5 Example query for a class "stack" with two methods, one named "push" with return type "void" and one named "pop" with return type "int"

```
{
    "_shards": {"failed": 0, "successful": 1, "total": 1},
    "hits": {
        "hits": [
            {"_id": "lintool/Cloud9/.../StackOfInts.java", ...,
                "_index": "AGORA", "_score": 15.214152, "_source": ...},
            {"_id": "CyanogenMod/android_frame.../.../Stack.java", ...,
                "_index": "AGORA", "_score": 14.03837, "_source": ...},
            {"_id": "android/platform_frameworks_base/.../Stack.java", ...,
                "_index": "AGORA", "_score": 14.03837, "_source": ...},
            ...
        ],
        "max_score": 15.214152, "total": 412275
    },
    "timed_out": false, "took": 109
}
```

Fig. 5.6 Example response for the query of Fig. 5.5

```
{
    "query": {"bool": {"should": [
        {"match": {"files.code.class.name": "Export"}},
        {"match": {"files.code.class.extends": "WizardPage"}},
        {"match": {"files.code.imports": "eclipse"}}
    ]}}
}
```

Fig. 5.7 Example query for a class with name containing "Export", which extends a class named "WizardPage" and the Java file has at least one import with the word "eclipse" in it

the results and select the one that fits best his/her scenario. Note that Elasticsearch documents also have a special field named "_source", which contains all the fields of the document. So, one can iterate over the results and access the source code of each file by accessing the field "source.content".

5.4.2.2   Finding Out How to Write Code in Context

Another scenario that is often handled using RSSEs [23] is that of finding out how to write code given some context. This scenario occurs when the developer wants to use a specific library or extend some functionality and wants to find relevant examples. An example query for this scenario is shown in Fig. 5.7.

In this case, the developer wants to find out how to create a page for an Eclipse wizard in order to export some element of his/her plugin project. Given the Eclipse Plugin API, it is easy to see that wizard pages extend the "WizardPage" class. So the query includes classes with name containing the word "Export" that extend the "WizardPage" class, while also containing "eclipse" imports. The results of this query are shown in Fig. 5.8.

The query retrieved 16681 documents in 16 ms. Note that not all results conform to all the criteria of the query, e.g., a result may extend "WizardPage", and its name may not contain the word "Export". However, the results that conform to most criteria appear at the top positions.

5.4.2.3   Searching for Patterns in Projects

The third scenario demonstrated is a particularly novel scenario in the area of CSEs. Often developers would like to find similar projects to theirs in order to understand how other developer teams (probably experienced ones or ones involved in large-scale projects) structure their code. Performing a generalization, one could argue that structure can be interpreted as a design pattern. In the context of a search, however,

```
{
    "_shards": {"failed": 0, "successful": 1, "total": 1},
    "hits": {
        "hits": [
            {"_id": "erlide/erlide/.../EdocExportWizardPage.java", ...,
                "_index": "AGORA", "_score": 11.272949, "_source": ...},
            {"_id": "rssowl/RSSOwl/.../ExportElementsPage.java", ...,
                "_index": "AGORA", "_score": 11.06465, "_source": ...},
            {"_id": "rssowl/RSSOwl/.../ExportOptionsPage.java", ...,
                "_index": "AGORA", "_score": 11.032191, "_source": ...},

            ...
        ],
        "max_score": 11.272949, "total": 16681
    },
    "timed_out": false, "took": 16
}
```

Fig. 5.8 Example response for the query of Fig. 5.7

it could also mean a simple way to structure one's code. Such a query is shown in Fig. 5.9.

The query is formed using the "has_child" functionality of Elasticsearch so that the returned projects have (at least) one class that implements an interface with name containing "Model", and another class that implements an interface with name similar to "View", and finally a third one that implements an interface with name containing "Controller". In short, projects following the Model-View-Controller (MVC) architecture are sought. Additionally, the query requests that the projects also have classes that extend "JFrame", so that they use the known GUI library swing. In fact, this query is quite reasonable since GUI frameworks are known to follow such architectures. The response for this query is shown in Fig. 5.10.

The query took 16 ms and returned 679 project documents. The files of these projects can be found using the "_parent" field of the files mapping.

5.4.2.4 Searching for Snippets

The fourth scenario demonstrates snippet search. Although syntax-aware search yields satisfactory results in class or method level, when it comes to raw code, searching for a snippet can also be helpful. An example query is shown in Fig. 5.11 and the corresponding response is shown in Fig. 5.12.

The syntax of the query includes a single match query on "analyzedcontent". The snippet is given as the value of the query where line feeds are represented using the

```
{
    "query": {"bool": {"should": [
        {"has_child": {"type": "files",
            "query": {"match": {"code.class.implements": "Model"}}
        }},
        {"has_child": {"type": "files",
            "query": {"match": {"code.class.implements": "View"}}
        }},
        {"has_child": {"type": "files",
            "query": {"match": {"code.class.implements": "Controller"}}
        }},
        {"has_child": {"type": "files",
            "query": {"match": {"code.class.extends": "JFrame"}}
        }}
    ]}}
}
```

Fig. 5.9 Example query for project with files extending classes "Model", "View", and "Controller" and also files extending the "JFrame" class

```
{
    "_shards": {"failed": 0, "successful": 1, "total": 1},
    "hits": {
        "hits": [
            {"_id": "Prototik/HoloEverywhere", ...,
                "_index": "AGORA", "_score": 2.0, "_source": ...},
            {"_id": "xamarin/XobotOS", ...,
                "_index": "AGORA", "_score": 2.0, "_source": ...},
            {"_id": "android/platform_frameworks_base", ...,
                "_index": "AGORA", "_score": 2.0, "_source": ...},

            ...

        ],
        "max_score": 2.0, "total": 679
    },
    "timed_out": false, "took": 16
}
```

Fig. 5.10 Example response for the query of Fig. 5.9

```
{
  "query": {
    "match": {
      "analyzedcontent": "File myfile = new File(\"myfile.xml\");\n
        DocumentBuilderFactory dbFactory = DocumentBuilderFactory.
                                                    newInstance();\n
        DocumentBuilder dBuilder = dbFactory.newDocumentBuilder();\n
        Document doc = dBuilder.parse(myfile);"
    }
  }
}
```

Fig. 5.11 Example query for a snippet that can be used to read XML files

```
{
  "_shards": {"failed": 0, "successful": 1, "total": 1},
  "hits": {
    "hits": [
      {"_id": "facebook/buck/.../WorkspaceGeneratorTest.java", ...,
        "_index": "AGORA", "_score": 1.8735983, "_source": ...},
      {"_id": "rhuss/jolokia/.../SpringConfigTest.java", ...,
        "_index": "AGORA", "_score": 1.363394, "_source": ...},
      {"_id": "openhab/openhab/.../DeviceCategoryLoader.java", ...,
        "_index": "AGORA", "_score": 1.2306316, "_source": ...},

      ...
    ],
    "max_score": 1.8735983, "total": 522327
  },
  "timed_out": false, "took": 78
}
```

Fig. 5.12 Example response for the query of Fig. 5.11

"\n" character. This particular example is a snippet for reading XML files. Possibly, the developer would want to know if the code is preferable or how others perform XML parsing.

This query took 78 ms and returned several documents (522327). Obviously not all of these perform XML parsing; several matches may include only the first line of the query. In any case, the most similar ones to the query are the ones with the largest "_score" at the top of the list.

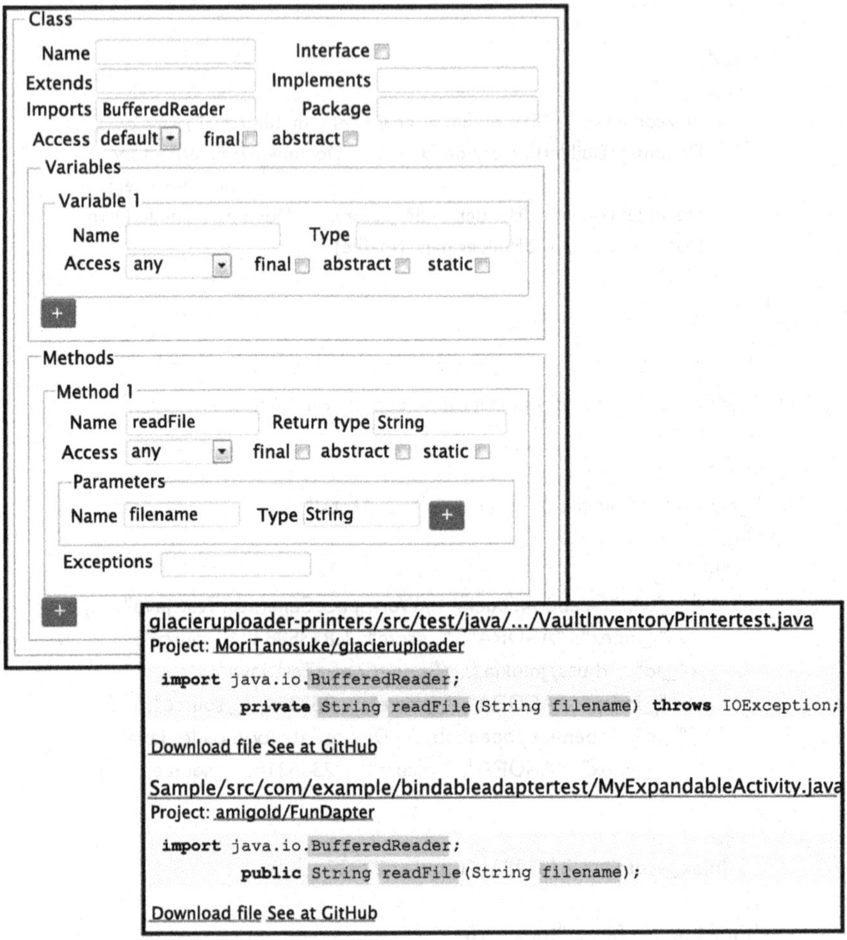

Fig. 5.13 Query for a file reader on the Advanced Search page of AGORA

5.4.3 Example Usage Scenario

We provide an example of using AGORA to construct three connected components: (a) one for reading texts from different files, (b) one for comparing extracted texts using an edit distance algorithm, and (c) one for storing the results in a data structure.

Figure 5.13 depicts a query for a component that reads a file into a string, including the first two results.

Defining a method "readFile" that receives a "filename" and returns a "String" is straightforward. The developer may optionally choose a file reading mechanism, by setting the import field. In Fig. 5.13, the developer has chosen "BufferedReader";

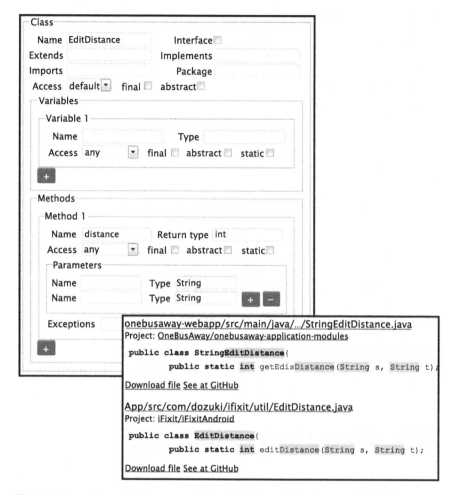

Fig. 5.14 Query for an edit distance method on the Advanced Search page of AGORA

however, the field could be set to "Scanner" or any other reader, or to "apache" to use the Apache file utilities, or even to "eclipse" to read the files as Eclipse resources.

Upon having read a collection of strings, we use the query of Fig. 5.14 to find a component that receives two strings and computes their edit distance. In this case, the developer searches for an "EditDistance" component that would return the edit distance of two strings (of which the variable names are not given since they are not necessary) as an integer. Similar results are obtained by setting "EditDistance" as a method (or even by searching for it in the simple search page of Fig. 5.4).

Storing the computed edit distances requires a pair that would contain the two strings (or possibly two string ids) and a Java HashMap that would have the pair as the key and the edit distance of the two strings as a value. A query for a class "Pair" with two variables "first" and "second" is shown in Fig. 5.15. Furthermore,

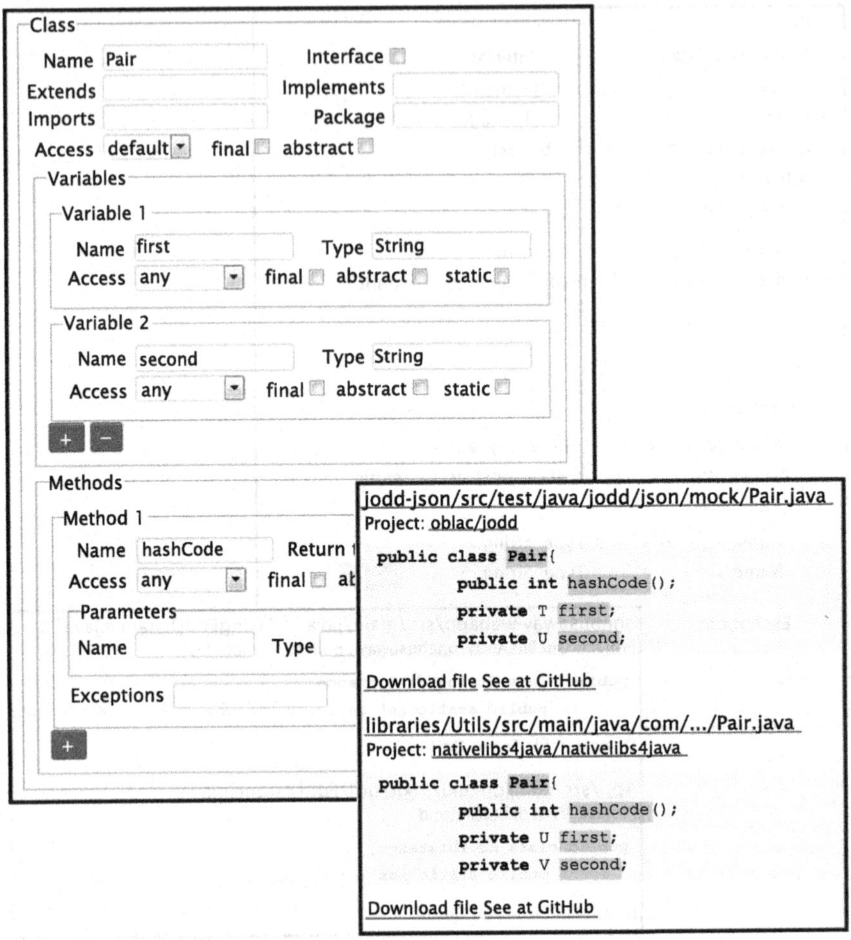

Class

Name Pair   Interface ☐
Extends   Implements
Imports   Package
Access default ▾   final ☐ abstract ☐

Variables

Variable 1

Name first   Type String
Access any   ▾   final ☐ abstract ☐ static ☐

Variable 2

Name second   Type String
Access any   ▾   final ☐ abstract ☐ static ☐

[+] [−]

Methods

Method 1

Name hashCode   Return
Access any   ▾   final ☐ ab

Parameters

Name   Type

Exceptions

[+]

jodd-json/src/test/java/jodd/json/mock/Pair.java
Project: oblac/jodd

```java
public class Pair{
        public int hashCode();
        private T first;
        private U second;
```

Download file See at GitHub

libraries/Utils/src/main/java/com/.../Pair.java
Project: nativelibs4java/nativelibs4java

```java
public class Pair{
        public int hashCode();
        private U first;
        private V second;
```

Download file See at GitHub

Fig. 5.15 Query for a pair class on the Advanced Search page of AGORA

given that the pairs should be HashMap keys, the query involves implementing the "hashCode" method.

The scenario of this section illustrates how the developer can use AGORA to find reusable components and thus reduce the effort required for his/her project. Similar use cases can be composed for the other types of AGORA queries (e.g., the ones described in the previous subsection).

5.5 Evaluation

Although the high-level comparison of AGORA with other CSEs is interesting (see Sect. 5.2.3), we need a quantitative measure to assess the effectiveness of our CSE. The measure of choice should reflect the relevance of the results retrieved by the engine. Thus, we need to examine whether the results for a query accurately cover the desired functionality. Therefore, in this section, we evaluate AGORA against two known CSEs: GitHub Search and BlackDuck Open Hub. We selected these two CSEs since they are representative of simple search and advanced search functionalities. In specific, BlackDuck supports searching using syntax-aware queries, while GitHub is used with text search. GitHub was actually chosen over all other simple search engines (like Searchcode) because of the size of its index. Most other engines of Table 5.1 are either not operational (e.g., Merobase) or provided for local installations (e.g., Krugle).

We evaluate the three CSEs on a simple component reuse scenario, as analyzed in the previous section. Typically, the developer wanting to find a specific component will submit the relevant query, and the CSE will return several results. Then, the developer would have to examine these results and determine whether they are relevant and applicable to his/her code. In our case, however, relying on examining the results for the CSEs ourselves would pose serious threats to validity. Instead, we determine whether a result is relevant and usable by using test cases (a logic also used in the next chapter). Our evaluation mechanism and the used dataset are presented in Sect. 5.5.1, while the results for the three CSEs are presented in Sect. 5.5.2.

5.5.1 Evaluation Mechanism and Dataset

Our dataset, which is shown in Table 5.6, comprises 13 components of variable complexity.

Note also that there are two versions of the "Stack" component to assess the effectiveness of the CSEs for different method names. For each of the components of Table 5.6, we have constructed queries for the CSEs and test cases to determine the functionality of the results. Concerning query construction, the query for AGORA is similar to the ones defined in the previous section. Since the GitHub Search API does not support syntax-aware search, we constructed a query that contains all methods of the component one after another. If this query returned no results, we created additional queries without the method return types, without the method parameters, and subsequently by including tokenized method and class names. An example sequence of queries is shown in Fig. 5.16.

The first query was adequate in cases where the method names were as expected, while subsequent queries were needed for more ambiguously defined names (such as the modified "Stack" component in Table 5.6). Concerning BlackDuck, it was easy

Table 5.6 Evaluation dataset for code search engines

Class	Methods
Account	deposit, withdraw, getBalance
Article	setId, getId, setName, getName, setPrice, getPrice
Calculator	add, subtract, divide, multiply
ComplexNumber	ComplexNumber, add, getRealPart, getImaginaryPart
CreditCardValidator	isValid
Customer	setAddress, getAddress
Gcd	gcd
Md5	md5Sum
Mortgage	setRate, setPrincipal, setYears, getMonthlyPayment
Movie	Movie, getTitle
Spreadsheet	put, get
Stack	push, pop
Stack2	pushObject, popObject

Query 1: void setAddress(String) String getAddress()

Query 2: setAddress(String) String getAddress()

Query 3: setAddress getAddress

Query 4: set address get address

Query 5: customer

Fig. 5.16 Example sequence of queries for GitHub

to construct the queries in the webpage of the service; however, it did not support any API. Therefore, we did it by hand and downloaded the first results for each query.

Upon constructing the queries for the CSEs we also created test cases for each component. In each test case, for a specific component, the names of the objects, variables, and methods are initially written with respect to the component. When a specific result is retrieved from a CSE, it is assigned a Java package, and the code is saved in a Java class file. After that, the test case is modified to reflect the class, variable, and method names of the class file. A modified test case for a class "Customer1" with functions "setTheAddress" and "getTheAddress", which is a result of the "Customer" component query is shown in Fig. 5.17. Variable "m_index" controls the order of the classes in the file. Ordering the methods of the class file in the same order as the test case (a procedure that, of course, does not change the functionality of the class file) can be performed by iterating over all possible combinations of them.

```
package package1;
import static org.junit.Assert.assertEquals;
import org.junit.Test;
import java.lang.reflect.*;

public class CustomerTest{
  @Test
  public void testAddress() throws Exception{
    int m_index = 0;
    Class<?> clazz = Class.forName("package1.Customer1");
    Object c = clazz.newInstance();
    Method m = clazz.getDeclaredMethod("setTheAddress", String.class);
    m.invoke(c,"test");
    m_index = 1;
    m = clazz.getDeclaredMethod("getTheAddress");
    assertEquals("Wrong string!","test",m.invoke(c));
  }
}
```

Fig. 5.17 Example modified test case for the result of a code search engine

Hence, our evaluation framework determines whether each result returned from a CSE is compilable and whether it passes the test. Note that compilability is also evaluated with respect to the test case, thus ensuring that the returned result is functionally equivalent to the desiderata of the developer. This suggests that the developer would be able to make adjustments to any results that do not pass the test.

5.5.2 Evaluation Results

In this subsection, we present the results of our evaluation for the components described in Sect. 5.5.1. Table 5.7 presents the number of results for each query, how many of them are compilable and how many passed the tests, for each CSE. We kept the first 30 results for each CSE, since keeping more results did not improve the compilability and test passing score for any CSE, and 30 results are usually more than a developer would examine in such a scenario.

Both AGORA and GitHub returned at least 30 results for each query. BlackDuck, on the other hand, did not return any results for three queries, while there are also two more queries for which the returned results are less than 10. If we examine the queries that returned few results, we conclude that it is due to differently defined method names. For example, although BlackDuck returned numerous results for a "Stack" with "push" and "pop" methods, it returned no results for a "Stack" with "pushObject" and "popObject". This is actually one of the strong points of AGORA,

Table 5.7 Compiled results and results that passed the tests for the three code search engines, for each query and on average, where the format for each value is number of tested results/number of compiled results/total number of results

Query	Code search engines		
	GitHub	AGORA	BlackDuck
Account	0/10/30	1/4/30	2/9/30
Article	1/6/30	0/5/30	2/3/15
Calculator	0/5/30	0/5/30	1/3/22
ComplexNumber	2/2/30	2/12/30	1/1/3
CreditCardValidator	0/1/30	1/4/30	2/3/30
Customer	2/3/30	13/15/30	7/7/30
Gcd	1/3/30	5/9/30	12/16/30
Md5	3/7/30	2/10/30	0/0/0
Mortgage	0/6/30	0/4/30	0/0/0
Movie	1/3/30	2/2/30	3/7/30
Spreadsheet	0/1/30	1/2/30	0/0/6
Stack	0/3/30	7/8/30	8/12/30
Stack[(2)]	0/0/30	7/8/30	0/0/0
Average number of results	30.0	30.0	17.4
Average number of compiled results	3.846	6.769	4.692
Average number of tested results	0.769	3.154	2.923

which is, in fact, due to its CamelCase analyzer (see Sect. 5.3.2.3). GitHub also would return minimum results, if not for the multi-query scheme we defined in Sect. 5.5.1.

Concerning the compilability of the results, and subsequently their compatibility with the test cases, the results of AGORA are clearly better than the results of the other two CSEs. In specific, AGORA is the only engine that returns compilable results for all 13 queries, while on average it returns more than 6.5 compilable results per query. GitHub and BlackDuck return roughly 4 and 4.5 results per query, respectively, indicating that our CSE provides more useful results in most cases.

The number of results that pass the test cases for each CSE is also an indicator that AGORA is quite effective. Given the average number of these results, shown in Table 5.7, it seems that AGORA is more effective than BlackDuck, while GitHub is less effective than both. Although the difference between AGORA and BlackDuck does not seem as important as in the compilability scenario, it is important to note that AGORA contains much fewer projects than both other CSEs, and thus it is possible that certain queries do not have test-compliant results in the index of AGORA. Finally, the average number of compilable results and results that passed the tests are visualized in Fig. 5.18, where it is clear that AGORA is the most effective of the three CSEs.

Fig. 5.18 Evaluation
diagram depicting the
average number of
compilable and tested results
for the three code search
engines

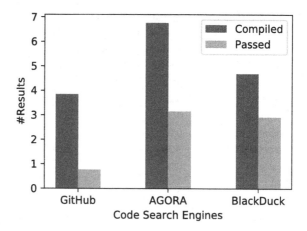

Upon examining whether the retrieved results for each CSE are useful for the component reuse scenario, it is also reasonable to determine whether these results are also ranked in an effective manner from each engine. For this reason, we use the search length as a metric for the ranking of a CSE. The search length is defined as the number of non-relevant results that the user must examine in order to find the first relevant result. One may also define subsequent values for the second relevant result, the third one, etc. In our case, we define two variants of this measure, one for the results that pass compilation and one for the results that pass the tests. Table 5.8 presents the search length for the two cases with regard to finding the first relevant (i.e., compilable or passed) result, for all the queries and for the three CSEs.

For the case of finding compilable results, AGORA clearly outperforms the other two CSEs, requiring on average to examine roughly 5 irrelevant results to find the first relevant one. By contrast, a search on GitHub or BlackDuck would require examining approximately 6.8 and 10 irrelevant results, respectively. The average search length for finding more than one compilable results is shown in Fig. 5.19a. In this figure, it is also clear that using AGORA requires examining fewer irrelevant results to find the first relevant ones. Given that a developer wants to find the first three compilable results, then on average he/she would have to examine 9.7 of the results of AGORA, 13.3 of the results of BlackDuck, and 16.5 of the results of GitHub.

In Table 5.8, the search length for the results that pass the tests is also shown for the three CSEs. In this case, BlackDuck seems to require examining fewer irrelevant results (ones failed the test) on average than AGORA, nearly 10.5 and 12.5, respectively, while GitHub is clearly outperformed requiring to examine more than 22 irrelevant results. However, it is important to note that the first relevant result of AGORA is included in the first 30 results for 10 out of 13 queries, whereas for BlackDuck for 9 out of 13 queries. It is also clear that the average of the search length for the CSEs is greatly influenced by extreme values. For example, BlackDuck is very effective for the "Account", "Calculator", and "ComplexNumber" queries. This is expected since we compare CSEs that contain different projects and subsequently

Table 5.8 Search length for the compiled and passed results for the three code search engines, for each result and on average, where the format for each value is passed/compiled

Query	Code search engines		
	GitHub	AGORA	BlackDuck
Account	0/30	2/2	0/0
Article	0/21	0/30	0/2
Calculator	0/30	8/30	0/0
ComplexNumber	0/0	0/0	0/0
CreditCardValidator	3/30	8/15	0/2
Customer	0/0	0/0	1/1
Gcd	25/25	1/8	5/5
Md5	1/9	0/0	30/30
Mortgage	0/30	4/30	30/30
Movie	23/23	21/21	0/1
Spreadsheet	4/30	20/23	30/30
Stack	2/30	1/2	4/4
Stack2	30/30	1/2	30/30
Average compiled	6.769	5.077	10.000
Average passed	22.154	12.538	10.385

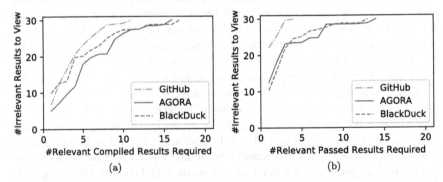

(a)   (b)

Fig. 5.19 Evaluation diagrams depicting **a** the search length for finding compilable results and **b** the search length for finding results that pass the tests, for the three code search engines

different source code components. Hence, AGORA is actually quite effective, considering that it contains less than 3000 projects, while GitHub and BlackDuck contain roughly 14 million and 650 thousand projects, respectively.

The average search length for finding more than one tested result is shown in Fig. 5.19b. As shown in this figure, both AGORA and BlackDuck are considerably more effective than GitHub. BlackDuck requires examining slightly fewer results for finding the first three results that pass the tests, whereas AGORA is clearly better when more results are required.

5.6 Conclusion

In this chapter, we have presented AGORA, a CSE with advanced searching capabilities, a comprehensive REST API and seamless integration with GitHub. The strength of AGORA lies in its architecture and its powerful indexing mechanism. We have evaluated AGORA against other CSEs and we have shown how it can be used for finding reusable components, using an example usage scenario. In the provided scenario, the developer would only need to issue three queries to AGORA, and then he/she would be presented with ready-to-use components (i.e., the ones ranked high in the result set). Compared to building the required components from scratch, using AGORA would save valuable development time and effort.

Concerning future work, we consider conducting developer studies in realistic scenarios to determine whether AGORA facilitates development in a reuse context and whether the features of AGORA are well received by the developer community. The index could also be extended by including quality metrics (e.g., static analysis metrics) for each file. Finally, we may employ query expansion strategies and/or incorporate semantics to further improve the relevance and/or the presentation of the retrieved results.

References

1. Thummalapenta S, Xie T (2007) PARSEWeb: a programmer assistant for reusing open source code on the web. In: Proceedings of the 22nd IEEE/ACM international conference on automated software engineering, ASE '07, New York, NY, USA. ACM, pp 204–213
2. Xie T, Pei J (2006) MAPO: mining API usages from open source repositories. In: Proceedings of the 2006 international workshop on mining software repositories, MSR '06, New York, NY, USA. ACM, pp 54–57
3. Hummel O, Janjic W, Atkinson C (2008) Code conjurer: pulling reusable software out of thin air. IEEE Softw 25(5):45–52
4. Lazzarini Lemos OA, Bajracharya SK, Ossher J (2007) CodeGenie: a tool for test-driven source code search. In: Companion to the 22nd ACM SIGPLAN conference on object-oriented programming systems and applications companion, OOPSLA '07, New York, NY, USA. ACM, pp 917–918
5. Diamantopoulos T, Symeonidis AL (2018) AGORA: a search engine for source code reuse. SoftwareX, page under review
6. Janjic W, Hummel O, Schumacher M, Atkinson C (2013) An unabridged source code dataset for research in software reuse. In: Proceedings of the 10th working conference on mining software repositories, MSR '13, Piscataway, NJ, USA. IEEE Press, pp 339–342
7. Linstead E, Bajracharya S, Ngo T, Rigor P, Lopes C, Baldi P (2009) Sourcerer: mining and searching internet-scale software repositories. Data Min Knowl Discov 18(2):300–336
8. Elasticsearch: RESTful, distributed search & analytics (2016). https://www.elastic.co/products/elasticsearch. Accessed April 2016
9. GitHub API, GitHub Developer (2016). https://developer.github.com/v3/. Accessed April 2016
10. Java Compiler Tree API (2016). http://docs.oracle.com/javase/8/docs/jdk/api/javac/tree/index.html. Accessed April 2016
11. Unicode Standard Annex #29 (2016) Unicode text segmentation. In: Davis M (ed) An integral part of The Unicode Standard. http://www.unicode.org/reports/tr29/. Accessed April 2016

12. CamelCase tokenizer, pattern analyzer, analyzers, Elasticsearch analysis (2016). http://www.elastic.co/guide/en/elasticsearch/reference/current/analysis-pattern-analyzer.html#_camelcase_tokenizer. Accessed April 2016
13. Lucene's practical scoring function, controlling relevance, search in depth, elasticsearch: The definitive guide (2016). http://www.elastic.co/guide/en/elasticsearch/guide/current/practical-scoring-function.html. Accessed April 2016
14. Manning CD, Raghavan P, Schütze H (2008) Introduction to information retrieval. Cambridge University Press, New York
15. Gamma E, Vlissides J, Johnson R, Helm R (1998) Design patterns: elements of reusable object-oriented software. Addison-Wesley Longman Publishing Co. Inc, Boston
16. Papamichail M, Diamantopoulos T, Symeonidis AL (2016) User-perceived source code quality estimation based on static analysis metrics. In: Proceedings of the 2016 IEEE international conference on software quality, reliability and security, QRS, Vienna, Austria, pp 100–107
17. Aggarwal K, Hindle A, Stroulia E (2014) Co-evolution of project documentation and popularity within github. In: Proceedings of the 11th working conference on mining software repositories, MSR 2014, New York, NY, USA. ACM, pp 360–363
18. Weber S, Luo J (2014) What makes an open source code popular on GitHub? In: 2014 IEEE international conference on data mining workshop, ICDMW, pp 851–855
19. Borges H, Hora A, Valente MT (2016) Understanding the factors that impact the popularity of GitHub repositories. In: 2016 IEEE international conference on software maintenance and evolution (ICSME), ICSME, pp 334–344
20. Dimaridou V, Kyprianidis A-C, Papamichail M, Diamantopoulos T, Symeonidis A (2017) Towards modeling the user-perceived quality of source code using static analysis metrics. In: Proceedings of the 12th international conference on software technologies - volume 1, ICSOFT, Setubal, Portugal, 2017. INSTICC, SciTePress, pp 73–84
21. Diamantopoulos T, Thomopoulos K, Symeonidis AL (2016) QualBoa: reusability-aware recommendations of source code components. In: Proceedings of the IEEE/ACM 13th working conference on mining software repositories, MSR '16, pp 488–491
22. Reiss SP (2009) Semantics-based code search. In: Proceedings of the 31st international conference on software engineering, ICSE '09, Washington, DC, USA. IEEE Computer Society, pp 243–253
23. Sahavechaphan N, Claypool K (2006) XSnippet: mining for sample code. SIGPLAN Not. 41(10):413–430

Chapter 6
Mining Source Code for Component Reuse

6.1 Overview

As already mentioned in the previous chapter, the open-source software development paradigm dictates harnessing online source code components to support component reuse. This way, the time and effort spent for software development is greatly reduced, while the quality of the projects is also notably improved. In this context, *Code Search Engines (CSEs)* have proven useful for locating source code components (see previous chapter). However, they also have their limitations. In specific, CSEs do not take into account the given context of a query, i.e., the source code and/or the functionality required by the developer. To overcome this shortcoming, several specialized systems have been developed in the broad area of *Recommendation Systems in Software Engineering (RSSEs)*. In the context of code reuse, RSSEs are intelligent systems, capable of extracting the query from the source code of the developer and processing the results of CSEs to recommend software components, accompanied by information that would help the developer understand and reuse the code. Lately, following the paradigm of *Test-Driven Development (TDD)*, several code reuse RSSEs further employ testing methods in order to determine whether the retrieved components have the desired functionality.

Although several of these *Test-Driven Reuse (TDR)* systems have been developed, their effectiveness when it comes down to realistic component search scenarios is arguable. As noted by Walker [1], most of these systems focus on the information retrieval problem of finding relevant software artifacts, while ignoring the syntax and the semantics of the source code. Additionally, according to Nurolahzade et al. [2], current RSSEs depend on matching signatures between components (i.e., methods with their parameters or fields), while limiting their retrieved results due to inflexible matching methods (i.e., similar methods with different names may not be matched) and/or absence of semantics. Even when results are returned, no additional postprocessing is performed, so extra effort is required for the developer to adapt them to his/her source code. Other important time-consuming challenges involve understanding the functionality of retrieved components and pointing out

© Springer Nature Switzerland AG 2020
T. Diamantopoulos and A. L. Symeonidis, *Mining Software Engineering Data for Software Reuse*, Advanced Information and Knowledge Processing,
https://doi.org/10.1007/978-3-030-30106-4_6

possible dependencies. Additionally, several RSSEs depend on static, non-updatable and sometimes obsolete software repositories, thus the quality of their results is arguable. Finally, another concern that can be raised involves the complexity of certain solutions, which do not return results in reasonable time.

In this chapter, we present *Mantissa*,[1] an RSSE that allows searching for software components in online repositories. Mantissa extracts the query from the source code of the developer and employs AGORA and GitHub to search for available source code. The retrieved results are ranked using a *Vector Space Model (VSM)*, while various *Information Retrieval (IR)* techniques and heuristics are employed in order to best rank the results. Additional source code transformations are performed so that each result can be ready-to-use by the developer. Finally, apart from relevance scoring, each result is presented along with useful information indicating its original source code, its control flow, as well as any external dependencies.

6.2 Background on Recommendation Systems in Software Engineering

6.2.1 Overview

The recommendation of McIlroy [3] for component reuse in the field of software engineering, dated back in 1968, is more accurate than ever. Living in a period where vast amounts of data are accessible through the Internet, and the time to market plays a vital role in industry, software reuse can result in time-saving, as well as to the development of high-quality and reliable software.

Aiming to facilitate code reuse, RSSEs have made their appearance during the last years [4, 5]. According to Robillard et al. [6], an RSSE is "a software application that provides information items estimated to be valuable for a software engineering task in a given context". In our case, the "software engineering task" refers to finding useful software components, while the "given context" is the source code of the developer.[2]

As stated by Walker [1], "recommendations for a software engineer play much the same role as recommendations for any user". In the case of software engineering, we could say that a general concept for a recommendation is as follows: an engineer would like to solve a problem for which there are several possible solutions available. The RSSE identifies the problem (extracting useful information from the given source code), decides on relevant results according to some metrics, and presents them to the engineer, in an appropriate manner.

---

[1]The term "Mantissa" refers to a sacred woman in ancient Greece, who interpreted the signs sent by the gods and gave oracles. In a similar way, our system interprets the query of the user and provides suitable results.

[2]See [7] for a systematic literature review on software engineering tasks that are facilitated using recommendation systems.

Although there are several code reuse recommendation systems, their functional steps are more or less similar. In specific, many contemporary code reuse recommendation systems [8–14] follow a five-stage process:

- Stage 1—Data Collection: Local repositories, CSEs, or even open-source code repositories are used for source code retrieval.
- Stage 2—Data Preprocessing: Upon extracting information from the retrieved source code, data are transformed in some representation so that they can later be mined.
- Stage 3—Data Analysis and Mining: Involves all the techniques used for extracting useful information for the retrieved components. Common techniques in this stage include clone detection, frequent sequences mining, etc.
- Stage 4—Data Postprocessing: Determines the recommendations of the system, usually in the form of a ranking of possible solutions.
- Stage 5—Results Presentation: The results are presented along with a relevance score and metrics concerning the functional or non-functional aspects of each result.

In the following subsections, we discuss the main approaches that have grown to be popular in the RSSE community.

6.2.2 Code Reuse Systems

One of the first code reuse RSSEs is CodeFinder, a system developed by Heninger [15]. CodeFinder does not make use of any special query languages; it is based on keywords, category labels, and attributes to conduct keyword matching. Its methodology is based on *associative networks* and is similar to that of the *PageRank* algorithm [16] that was published a few years later. CodeWeb [17] is another early developed system for the KDE application framework that takes into account the structure of a query to recommend results, which are accompanied by documentation. Although it is one of the first RSSEs to use *association rules*, it is actually a browsing system rather than a search engine, thus limiting its results.

A promising system named CodeBroker was published in 2002 by Ye and Fischer [18]. CodeBroker was developed as an add-on for the Emacs editor. The tool forms the query from documentation comments or signatures and is the first to utilize a local repository using Javadoc comments. CodeBroker was also largely based on user feedback, enabling retrieval by reformulation, while it also presented results accompanied by evaluation metrics (e.g., *recall* and *precision*) and documentation. However, the use of only a local database and its dependency on Javadoc comments restricted the search results of CodeBroker.

From 2004 onward, when the Eclipse Foundation[3] was established, many RSSEs were developed as plugins for the Eclipse IDE. Two of the first such systems are

---

[3]https://eclipse.org/org/foundation/.

Strathcona [19] and Prospector [20], both published in 2005. Strathcona extracts queries from code snippets and performs structure matching, promoting the most frequent results to the user, including also UML diagrams. However, it also uses a local repository, thus the results are limited. Prospector introduced several novel features, including the ability to generate code. The main scenario of this tool involves the problem of finding a path between an input and an output object in source code. Such paths are called *jungloids* and the resulting program flow is called a jungloid graph. Prospector also requires maintaining a local database, while the main drawback of the tool is that API information may be incomplete or even irrelevant, so the results are limited.

By 2006, the first CSEs made their appearance, and most RSSEs started to rely on them for finding source code, thus ensuring that their results are up-to-date. One of the first systems to employ a CSE was MAPO [10], a system developed by Xie and Pei, with a structure that resembles most modern RSSEs. Upon downloading source code from open-source repositories, MAPO analyzes the retrieved snippets and extracts sequences from the code. These sequences are then mined to provide a ranking according to the query of the developer. XSnippet, published in 2006 by Sahavechaphan and Claypool [8], is another popular RSSE which is similar to MAPO. However, it uses structures such as acyclic graphs, B+ trees, and an extension of the *Breadth-First Search (BFS)* algorithm for mining.

PARSEWeb is another quite popular system developed by Thummalapenta and Xie [9]. Similarly to MAPO, PARSEWeb recommends API usages using code snippets from the Google Code Search Engine.[4] The service uses both *Abstract Syntax Trees (AST)* and *Directed Acyclic Graphs (DAGs)* as source code representations and clusters similar sequences. PARSEWeb employs heuristics for extracting sequences, while it also enables query splitting in case no adequate results have been found. All aforementioned systems are similar in terms of data flow. Considering the architecture of PARSEWeb (shown in Fig. 6.1), a typical RSSE has a downloader that receives the query of the user and retrieves relevant code (phase 1). After that, a code analyzer extracts representations that are then mined to produce the results (phase 2), which, before presented to the user, often undergo some postprocessing (phase 3). In the case of PARSEWeb, the system extracts method invocation sequences, while the mining part (sequence postprocessor) clusters the sequences. If the clustering process produces very few sample sequences, PARSEWeb also allows splitting the query and issuing two consecutive queries to the system.

Lately, several of the systems that have been developed (such as Portfolio [21], Exemplar [22], or Snipmatch [23], the snippet retrieval recommender of the Eclipse IDE) accept queries in natural language. Bing Code Search [11] further introduces a multi-parameter ranking system, which uses source code and semantics originating from results of the Bing search engine. Finally, certain RSSEs have harnessed the power of question-answering systems for recommending source code, such as Example Overflow [24] which recommends Stack Overflow snippets, while there is

---

[4]As already mentioned in the previous chapter, the Google Code Search Engine was discontinued in 2013.

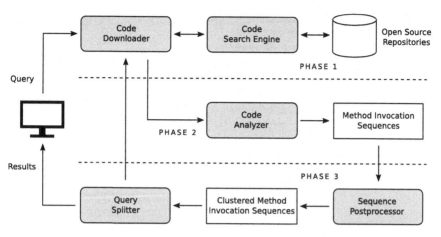

Fig. 6.1 Architecture of PARSEWeb [9]

also a preference in dynamic and interactive systems, such as the recently developed CodeHint [12].

6.2.3 Test-Driven Reuse Systems

As already discussed, in an attempt to ensure that the desired functionality is covered, several RSSEs assess the retrieved components using test cases. The concept of TDD is based on writing test cases before writing the actual code. Additionally, the code to be added has to be the minimum possible required to pass the test, in order to avoid any duplication [25]. Thus, the goal is to produce clean code that exhibits the required functionality. TDD combines three individual activities: *coding*, *testing*, and *refactoring*. A major advantage of this approach is that both the source code and the tests have been checked and work well; the tests fail before writing any piece of code and pass after adding the code, while the code is checked against these valid tests.

In the context of reuse, several researchers have attempted to exploit the advantages of TDD. Although the notion of TDR can be traced back to 2004 [26], most work on test-driven RSSEs for reuse has been conducted after 2007 [2].

One of the first systems to incorporate TDD features is Code Conjurer, an RSSE developed by Hummel et al. [13, 27, 28] as an Eclipse plugin. Code Conjurer uses the CSE Merobase, which has been created by the same research group [29] to support syntax-aware queries for finding reusable components. A screenshot of the tool is shown in Fig. 6.2. Using Code Conjurer, the developer needs only to write a test, such as the one at the top of Fig. 6.2. After that, the system retrieves relevant components (lower left of the figure), executes them along with test, and allows the developer to examine them (lower right of the figure) before adding them to the source code. The main advantages of Code Conjurer are its powerful heuristics, including

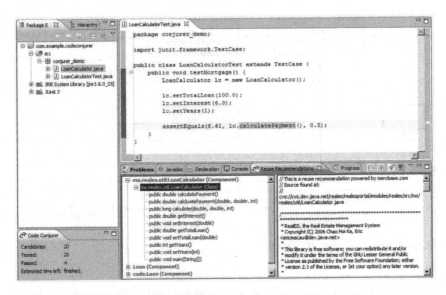

Fig. 6.2 Example screen of Code Conjurer for a loan calculator component [13]

clustering similar results and accompanying the results with complexity metrics. The
tool also supports a proactive mode, meaning that queries can be created dynamically
whenever the developer adds, modifies, or deletes a piece of code. Although Code
Conjurer is an effective RSSE, its matching mechanism is not flexible enough to deal
with small differences between the retrieved code and the tests, while the adaptation
(advanced) version of the tool is too computationally complex. Additionally, its CSE,
Merobase, has been discontinued, so the tool is not functional.

Another interesting approach is CodeGenie, implemented by Lemos et al. [14, 30,
31]. Similarly to Code Conjurer, CodeGenie employs its own CSE, Sourcerer [32,
33], in order to retrieve source code components. A major advantage of CodeGenie
is the ability to extract the query from the test code of the developer, thus the latter
has to write only a JUnit test. However, as pointed out in [34], the extraction of the
query from the test case could lead to failure as this is a non-trivial task. Additionally,
the tool does not detect duplicate results, while it also suffers from modifications in
the code of the components.

Reiss follows a slightly different concept in S6 [35, 36]. S6 is a web-based application that integrates several CSEs and repositories, including Krugle[5] and GitHub.
Although the developer has to compose more complex queries (e.g., semantic keywords), the retrieved results are refactored and simplified, making S6 possibly the
only RSSE that integrates the code refactoring aspect of TDD. An example screen of
S6 is shown in Fig. 6.3, where it is clear that the returned results are indeed quite useful; however, the tool offers some options that, at first sight, may be overwhelming.

---

[5]http://opensearch.krugle.org/.

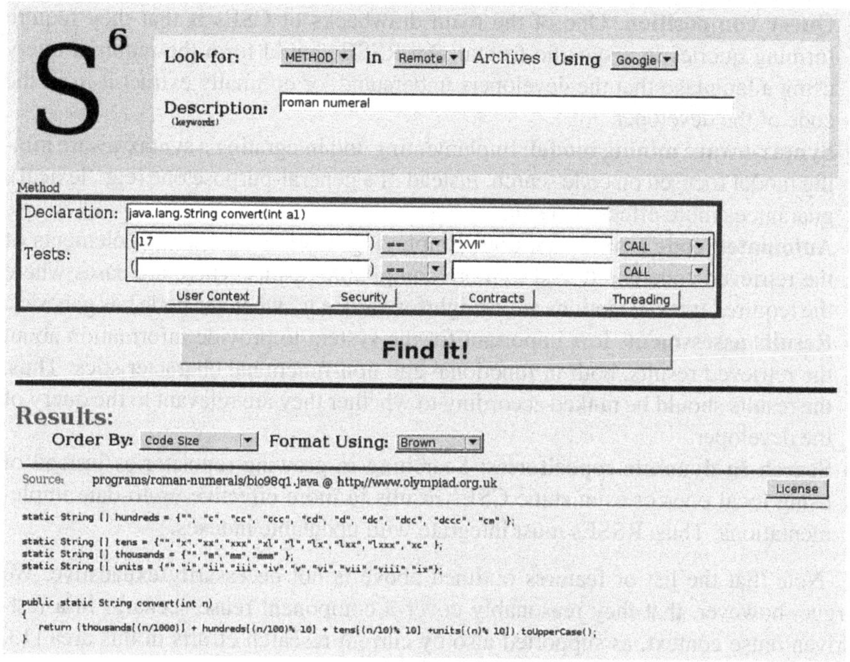

Fig. 6.3   Example screen of S6 for a converter of numbers to roman numerals [35]

Furthermore, the service often returns duplicates, while most results are incomplete snippets that confuse the developer, instead of helping him/her.

Another interesting system is FAST, an RSSE in the form of an Eclipse plugin developed by Krug [34]. The FAST plugin makes use of the Merobase CSE and emphasizes automated query extraction; the plugin extracts the signature from a given JUnit test and forms a query for the Merobase CSE. The retrieved results are then evaluated using automated testing techniques. A drawback of this system is that the developer has to examine the query extracted from the test before executing the search. Additionally, the service does not make use of growing software repositories, while it employs the deprecated CSE Merobase so it is also not functional.

6.2.4   Motivation

Although several RSSEs have been developed during the latest years, most of them have important limitations. For instance, not all are integrated with updatable software repositories, while their syntax-aware features are limited to exact matching between method and parameter names. We argue that an RSSE should exhibit the following features:

- **Query composition**: One of the main drawbacks of CSEs is that they require forming queries in a specific format. An RSSE should form the required query using a language that the developers understand, or optimally extract it from the code of the developer.
- **Syntax-aware mining model**: Implementing and integrating a syntax-aware mining model focused on code search, instead of a general-purpose one (e.g., lexical), guarantees more effective results.
- **Automated code transformation**: Employing heuristics to modify elements of the retrieved code could lead to more compilable results. There are cases where the required transformations are straightforward, e.g., when the code has generics.
- **Results assessment**: It is important for the system to provide information about the retrieved results, both in functional and non-functional characteristics. Thus, the results should be ranked according to whether they are relevant to the query of the developer.
- **Search in dynamic repositories**: Searching in growing repositories instead of using local ones or even static CSEs results in more effective up-to-date implementations. Thus, RSSEs must integrate with updatable indexes.

Note that the list of features outlined above is not necessarily exhaustive. We argue, however, that they reasonably cover a component reuse scenario in a test-driven reuse context, as supported also by current research efforts in this area [13, 14, 34, 35].

6.2.5   Mantissa Offerings

Upon presenting the current state of the art on TDR systems, we compare these systems to Mantissa with regard to the features defined in Sect. 6.2.4. The comparison, which is shown in Table 6.1, indicates that Mantissa covers all the required features/criteria, while the other systems are mostly limited to full coverage of two or three of these features.

Table 6.1 Feature comparison of popular test-driven reuse systems and Mantissa

Code reuse system	Feature 1	Feature 2	Feature 3	Feature 4	Feature 5
	Query composition	Syntax-awareness	Code transformation	Results assessment	Search in dynamic repos
Mantissa	✓	✓	✓	✓	✓
Code Conjurer	✓	✓	✓	✓	×
CodeGenie	✓	✓	×	✓	×
S6	×	✓	✓	✓	✓
FAST	✓	✓	×	✓	×

✓ Feature supported,   × Feature not supported,   ✓ Feature partially supported

At first, syntax-aware queries (feature 2) are supported by all systems; however, certain RSSEs, such as S6 and FAST, require the developer to construct the query in a specific format that he/she may not be familiar with. By contrast, Mantissa supports query composition (feature 1), as it extracts the query from the source code of the developer, thus requiring no additional training. Code Conjurer and CodeGenie are also quite effective for extracting the query of the developer; however, they do not fully support automated code transformations (feature 3), thus they may omit useful results. Mantissa, on the other hand, performs source code transformations in order to match as many results as possible to the query.

The assessment of the results is another crucial feature for code reuse RSSEs (feature 4). Concerning TDR systems, this feature is usually translated to assessing whether the results pass the provided test cases, or at least whether they can be compiled along with the test cases. Most TDR systems, including Mantissa, effectively cover this feature. Mantissa, however, further inspects the retrieved components to provide useful information indicating not only their origin (original files from GitHub repo), but also their control flow as well as any external dependencies. As a result, developers can easily understand the inner functionality of a component and integrate it into their source code. Another quite important feature of Mantissa is its integration with dynamically updated software repositories (feature 5). This ensures that the results are always up-to-date. As shown in Table 6.1, this feature is usually neglected by current TDR systems (except for S6). In summary, we argue that none of the reviewed systems and, to the best of our knowledge, no other system effectively covers the criteria defined in Sect. 6.2.4. In the following section, we present Mantissa, which supports these features and effectively retrieves reusable software components.

6.3 Mantissa: A New Test-Driven Reuse Recommendation System

In this section, we present Mantissa [37], a TDR system designed to support the features of Sect. 6.2.4.

6.3.1 Overview

The architecture/data flow of Mantissa is shown in Fig. 6.4. Mantissa comprises six concrete modules: the *Downloader*, the *Parser*, the *Miner*, the *Transformer*, the *Tester*, and the *Flow Extractor*. The developer provides the input query as a source code *signature*. A signature is quite similar to an interface for a component; it comprises all of its methods. Additionally, the developer has to provide the corresponding test for the component. Initially, the Parser receives the signature and parses it to ex-

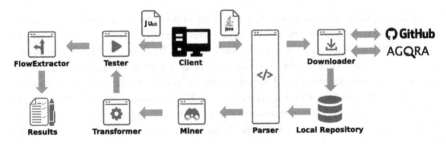

Fig. 6.4 The architecture/data flow of Mantissa

Fig. 6.5 Example signature
for a class "Stack" with
methods "push" and "pop"

```
public class Stack {
    public void push(Object o){}
    public Object pop(){}
}
```

tract its AST. The downloader uses the AST to compose the queries for the connected
CSEs. After that, it issues the queries, retrieves the results, and stores them in a local
repository.

The retrieved results are then analyzed by the Parser, which extracts the AST
for every result. The Miner ranks the results using a syntax-aware scoring model
based on information retrieval metrics. The retrieved files are then passed to the
transformer module, which performs source code transformations on them in order
to syntactically match them to the query. The tester compiles the files in order to
execute them against the given test cases. The ranked list of results involves the
score for each component, along with these elements, as well as information as to
whether the result is compilable and whether it has passed the tests. Finally, the
control flow and the dependencies are extracted for each result by the flow extractor,
while links to grepcode[6] are provided for each dependency/import.

Note that the architecture of Mantissa is language-agnostic, as it can retrieve functional software components written in different programming languages. In specific,
supporting a programming language requires developing a Parser module in order to
extract the ASTs of source code components, and a Tester module so as to test the
retrieved results. In the following paragraphs, we provide an implementation of the
modules of Mantissa for the Java programming language.

At first, the input signature of Mantissa is given in a form similar to a Java interface
and any test cases are written in JUnit format. An example signature for a Java "Stack"
component with two methods "push" and "pop" is shown in Fig. 6.5.

Note also that a wild character (_) can be used for cases where the user would
like to ignore the class/method name, while Java generic types can also be used.

---

[6]http://grepcode.com/.

Fig. 6.6 Example AST for the source code example of Fig. 6.5

```
<class>
    <name>Stack</name>
    <method>
        <name>push</name>
        <parameters>
            <type>Object</type>
        </parameters>
        <return>void</return>
    </method>
    <method>
        <name>pop</name>
        <return>Object</return>
    </method>
</class>
```

6.3.2 Parser

The parser module is used for extracting information from the source code of the developer as well as for extracting the ASTs of the retrieved results. The module employs srcML,[7] a tool which extracts the AST in XML format, where mark-up tags denote elements of the abstract syntax for the language. The tool also allows refactoring by transforming XML trees back to source code. The AST for a file includes all the elements that may be defined in the file: packages, imports, classes, methods, etc. An example of stripped-down (without modifiers) AST for the file of Fig. 6.5 is shown in Fig. 6.6.

6.3.3 Downloader

The downloader module is used to search in the integrated CSEs of the system, retrieve relevant files, and form a temporary local repository so that the files are further analyzed by the miner module. The module provides the developer with the ability to search either in Github Search or in our CSE AGORA that was analyzed in the previous chapter. We present the integration of both CSEs in the following subsection.

---

[7]http://www.srcml.org/.

Fig. 6.7 Example sequence
of queries for Searchcode

Query 1: Stack void push Object pop
Query 2: Stack void push
Query 3: Stack Object pop
Query 4: Stack
Query 5: void push Object pop
Query 6: void push
Query 7: Object pop

6.3.3.1   Integration with Code Search Engines

Concerning the integration of our system with GitHub, the latter provides an API for code search; however, it has two major limitations: (a) issuing a maximum number of requests per hour and (b) restricting its code search feature to specific repositories/users, thus not allowing to search in the whole index. For this reason, we decided to implement a custom downloader in order to overcome the latter limitation.

Our implementation uses the Searchcode CSE,[8] in order to initially find GitHub users whose repositories are possibly relevant to the query. In specific, the downloader uses the AST of the queried component in order to compose a number of queries for Searchcode. These queries are issued one after another until a specified number of results from different GitHub users are returned (tests have shown that retrieving the first 30 users is quite adequate). The downloader issues three types of queries:

- the name of the class and a combination of the methods, including their names and return types;
- the name of the class on its own as is and tokenized if it can be split into more than one tokens; and
- the combination of methods on their own, including the name and return type of each method.

Figure 6.7 depicts a list of queries for the "Stack" of Fig. 6.5.

Upon issuing the queries to Searchcode, the downloader identifies 30 user accounts that are a good fit for the query. Since, however, GitHub allows searching for more than 30 users, we decided to exploit this by also adding the most popular users to this list (popularity is determined given the number of stars of their repositories). In specific, GitHub allows performing queries in the source code of up to 900 user accounts per minute, i.e., 30 requests per minute each including 30 users.

Similarly to Searchcode, GitHub does not support regular expressions, hence the downloader formulates a list of queries in order to retrieve as many relevant results as possible. The list of queries is slightly different from what was used in the Searchcode CSE, since in the case of GitHub the required results are the files themselves and not only the repositories of GitHub users. In specific, the queries to GitHub include the following:

---

[8]https://searchcode.com/.

Fig. 6.8 Example sequence of queries for GitHub

Query 1: void push(Object) Object pop()
Query 2: push(Object) pop()
Query 3: push pop
Query 4: Stack

- the methods of the component, including their type, their name, and the type of their parameters;
- the names and the parameter types of the methods;
- the names of the methods;
- the tokenized names of the methods; and
- the tokenized name of the class.

The relevant example of a list of queries for the "Stack" component of Fig. 6.5 is shown in Fig. 6.8.

Finally, if the query is longer than 128 characters, then its terms are removed one at a time, starting from the last one until this limitation of GitHub is satisfied.

As indicated, GitHub imposes several restrictions on its API. Thus, we also use AGORA, which is better oriented toward component reuse. The syntax-aware search of AGORA allows finding Java objects with specific method names, parameter names, etc., while its API does not impose any limitations to the extent of the index to be searched. In our case, the downloader takes advantage of the syntax-aware features of AGORA to create queries such as the one shown in the searching for a reusable component scenario of AGORA (see Sect. 5.4.2.1).

6.3.3.2 Snippets Miner

For bandwidth reasons, most CSEs usually restrict the users with respect to the number of results that can be downloaded within a time window. For example, in the case of GitHub, downloading all the results one by one is rather ineffective. However, CSEs usually return a list of results including snippets that correspond to the matching elements of the query. To accelerate our system and ensure that a result is at least somewhat relevant before downloading it, we have constructed a *Snippets Miner* within the downloader module. The Snippets Miner component parses the snippets provided by the CSE and determines which are relevant to the query, ensuring the downloader downloads only these and ignores any remaining files.

Note that the Snippets Miner component can be adjusted to any CSE that returns snippets information. It employs syntax-aware features, thus in our case, it was not required to employ it for AGORA that already has these matching capabilities. The component, however, had to be activated for the results of GitHub.

At first, the component works by eliminating duplicates using the full filename of the result file (including the project name) as well as the *sha hash* returned by the

Fig. 6.9 Regular expression
for extracting method
declarations from snippets

`([\w\<\>\[\[]+)?\s+(\w+)?\((.*?)\)\s*(throws\s*\w+)*{`

    return type   name parameters   body

GitHub API for each result. After that, the component extracts method declarations
from each result and checks whether they comply with the original query. Note that
using the parser is not possible since the snippets are incomplete; thus, the method
declarations are exported using the regular expression of Fig. 6.9.

Upon extracting method declarations, the Snippets Miner performs matching between the methods that were extracted by the currently examined snippet and the
methods requested in the query of the developer. The matching mechanism involves
checking whether there are similar names and return types of methods, as well as
parameter names and types. For the return types, only exact matches are checked,
while for parameters the possible permutations are considered. A total score is given
to each result using the assigned score of each of these areas. The scoring scheme
used is explained in detail in Sect. 6.3.4, since it is the same as one of the miner
modules of Mantissa. The Snippets Miner is used to keep results that have score
greater than 0 and discard any remaining files. If, however, the retrieved results are
less than a threshold (in our case 100), then results that scored 0 are also downloaded
until that threshold is reached.

6.3.4   Miner

Upon forming a local repository with the retrieved results, the parser extracts their
ASTs as described in Sect. 6.3.2, while the miner ranks them according to the query of
the developer. The miner consists of three components: the *preprocessor*, the *scorer*,
and the *postprocessor*. These are analyzed in the following paragraphs.

6.3.4.1   Preprocessor

The first step before ranking the results involves detecting and removing any duplicate
files. There are several algorithms for detecting duplicate files while current literature
also includes various sophisticated code clone detection techniques. However, these
methods are usually not oriented toward code reuse; when searching for a software
component, finding similar ones is actually encouraged, what one would consider
excess information is *exact* duplicates. Thus, the preprocessor eliminates duplicates
using the MD5 algorithm [38] in order to ensure that the process is as fast as possible.
Initially, the files are scanned and an MD5 hash is created for each of them and after
that the hashes are compared to detect duplicates.

6.3.4.2 Scorer

The scorer is the core component of the miner module, which is assigned with scoring and ranking the files. It receives the ASTs of the downloaded files as input and computes the similarity between each of the results with the AST of the query. Our scoring methodology is similar to the *Vector Space Model (VSM)* [39], where the *documents* are source code files, while the *terms* are the values of the AST nodes (XML tags) as defined in Sect. 6.3.2. The scorer practically creates a vector for each result file and the query and compares these vectors.[9]

Since each component has a class, initially a class vector is created for each file. The class vector has the following form:

$$\vec{c} = [score(name), score(\vec{m}_1), score(\vec{m}_2), \ldots, score(\vec{m}_n)] \qquad (6.1)$$

where *name* is the name of the class and $\vec{m}_i$ is the vector of the ith method of the class, out of n methods in total. The score of the name is a metric of its similarity to the name of the requested component, while the score of each method vector is a metric of its similarity to the method vector of the requested component, which includes the name of the method, the return type, and the type of its parameters. A method vector $\vec{m}$ is defined as follows:

$$\vec{m} = [score(name), score(type), score(p_1), score(p_2), \ldots, score(p_m)] \quad (6.2)$$

where *name* is the name of the method, *type* is the return type of the method, and p_j is the type of the jth parameter of the method, out of m parameters in total.

Given Eqs. (6.1) and (6.2), we are able to represent the full signature of a component as a class vector. Thus, comparing two components comes down to comparing these vectors that contain numeric values in the range [0, 1]. The *score* functions are computed for each document-result file given the signature of the query. Computing the vector values of Eqs. (6.1) and (6.2) requires a function for computing the similarity between sets, for the methods and the parameters, and a function for computing the similarity between strings, for class names, method names and return types, and parameter types.

Note that using only a string matching method is not adequate, since the maximum score between two vectors such as the ones of Eqs. (6.1) and (6.2) also depends on the order of the elements. Methods, for example, are not guaranteed to have the same order in two files, as parameters do not necessarily have the same order in methods. Thus, the scorer has to find the best possible matching between two sets of methods or parameters, respectively. Finding the best possible matching between two sets is a problem similar to the *Stable Marriage Problem (SMP)* [40], where

---

[9]In specific, the main difference between our scoring model and the VSM is that our model is hierarchical, i.e., in our case a vector can be composed by other vectors. Our model has levels; at the class level it is practically a VSM for class vectors, while class vectors are further analyzed at the method level.

$SetsMatching(U, V)$
   $MatchedU = \{\}$
   $MatchedV = \{\}$
   $MatchedPairs = \{\}$
   $Pairs = \{(u, v, score(u, v))\,\forall u \in U, v \in V\}$
   Sort $Pairs$ in descending order according to their score
   for each $(u, v, score(u, v)) \in Pairs$:
      if $u \notin MatchedU$ and $v \notin MatchedV$
        $MatchedPairs = MatchedPairs \cup \{(u, v, score(u, v)))\}$
   return $MatchedPairs$

Fig. 6.10 Algorithm that computes the similarity between two sets of elements, U and V, and returns the best possible matching as a set of pairs $MatchedPairs$

two sets of elements have to be matched in order to maximize the profit given the preferences of the elements of the first set to the elements of the second and vice versa.[10] However, in our case, the problem is simpler, since the preference lists for both sets are symmetrical.

Hence, we have constructed an algorithm that finds the best possible matching among the elements of two sets. Figure 6.10 depicts the $SetsMatching$ algorithm.

The algorithm receives sets U and V as input and outputs the $MatchedPairs$, i.e., the best matching between the sets. At first, two sets are defined, $MatchedU$ and $MatchedV$, to keep track of the matched elements of the two initial sets. After that, all possible scores for the combinations of the elements of the two sets are computed and stored in the $Pairs$ set in the form $(u, v, score(u, v))$, where u and v are elements of the sets U and V, respectively. The $Pairs$ set is sorted in descending order according to the scores, and then the algorithm iterates over each pair of the set. For each pair, if neither of the elements is already present in the $MatchedU$ and $MatchedV$ sets, then the pair is matched and added to the $MatchedPairs$ set.

Finally, upon having confronted the problem of determining the matching between classes and methods, the only remaining issue is devising a string similarity scheme. The scheme shall be used to compute the similarity between Java names and types. Note that the string similarity scheme has to comply with the main characteristics of the Java language, e.g., handling camelCase strings, while at the same time employing NLP semantics (e.g., stemming). Thus, the first step before comparing two strings is preprocessing. Figure 6.11 depicts the preprocessing steps.

At first, the string is *tokenized* using a tokenizer that is consistent with the naming conventions of Java. The tokenizer splits the string into tokens on uppercase characters (for camelCase strings), numbers, and the underscore (_) character. Additionally,

---

[10] According to the original definition of the SMP, there are N men and N women, and every person ranks the members of the opposite sex in a strict order of preference. The problem is to find a matching between men and women so that there are no two people of opposite sex who would both rather be matched to each other than their current partners.

Fig. 6.11 String preprocessing steps

Table 6.2 String preprocessing examples

String	Tokenization	Stop words removal	Stemming
pushing_item	Pushing item	Pushing item	Push item
add2Numbers	add numbers	Add numbers	Add number
test1Me_Now	Test me now	Test now	Test now

it removes numbers and common symbols used in types ($<$ and $>$) and arrays ([and]) and converts the text to lowercase.

The next step involves removing *stop words* and terms with less than three characters, since they would not offer any value to the comparison. We used the stop words list of NLTK [41]. Finally, the PortStemmer of NLTK [41] is used to stem the words, applying both inflectional and derivational morphology heuristics. Table 6.2 illustrates these steps using different examples. The leftmost column shows the original string while subsequent columns show the result after performing each step, so the rightmost column shows the resulting tokens.

Finally, upon preprocessing the strings that are to be compared, we use string similarity methods to compute a score of how similar the strings are. For this, we used two different metrics: the *Levenshtein similarity* [42] and the *Jaccard index* [43]. The Levenshtein similarity is used by the scorer module in order to perform a strict matching that takes also the order of the strings into account, while the Jaccard index is used by the Snippets Miner component of the downloader (see Sect. 6.3.3.2) so that the matching is less strict.

As the Levenshtein distance is applied to strings, the token vectors are first merged. The *Levenshtein distance* between two strings is defined as the minimum number of character insertions, deletions, or replacements required to transform one string into the other. For our implementation, the cost of replacing a character was set to 2, while the costs of the insertion and deletion were set to 1. Then, for two strings, s_1 and s_2, their Levenshtein similarity score is computed as follows:

$$L(s_1, s_2) = 1 - \frac{distance(s_1, s_2)}{max\{|s_1|, |s_2|\}} \tag{6.3}$$

As shown in Eq. (6.3), the distance between the two strings is normalized in [0, 1] by dividing by the total number of characters in the two strings, and the result is subtracted from 1 to offer a metric of similarity.

The Jaccard index is used to compute the similarity between sets. Given two sets, U and V, their Jaccard index is the size of their intersection divided by the size of their union:

$$J(U, V) = \frac{|U \cap V|}{|U \cup V|} \tag{6.4}$$

In our case, the sets contain tokens, so the intersection of the two sets includes the tokens that exist in both sets, while their union includes all the tokens of both sets excluding duplicates.

Upon having constructed the algorithms and the string metrics required for creating the vector for each result, we now return to the VSM defined by Eqs. (6.1) and (6.2). Note that all metrics defined in the previous paragraphs are normalized, so all dimensions for the vector of a document will be in the range [0, 1]. The query vector is the all-ones vector. The final step is to compute a score for each document-result vector and rank these vectors according to their similarity to the query vector.

Since the Tanimoto coefficient [44] has been proven to be effective for comparing bit vectors, we have used its continuous definition for the scorer. Thus, the similarity between the query vector $\vec{Q}$ and a document vector $\vec{D}$ is computed as follows:

$$T(\vec{Q}, \vec{D}) = \frac{\vec{Q} \cdot \vec{D}}{||\vec{Q}||^2 + ||\vec{D}||^2 - \vec{Q} \cdot \vec{D}} \tag{6.5}$$

The nominator of Eq. (6.5) is the dot product of the two vectors while the denominator is the summation of their Euclidean norms minus their dot product. Finally, the scorer outputs the result files, along with their scores that indicate their relevance to the query of the developer.

As an example, we can use the signature of Fig. 6.5. The "Stack" component of this signature has two methods, "push" and "pop", so its vector would be an all-ones vector with three elements, one for the name of the class and one for each method:

$$\vec{c}_{Stack} = [1, 1, 1] \tag{6.6}$$

and the two vectors for "push" and "pop" would be

$$\vec{m}_{pop} = [1, 1] \tag{6.7}$$

and

$$\vec{m}_{pop} = [1, 1] \tag{6.8}$$

respectively, since "push" has a name, a return type ("void"), and a parameter type ("Object"), while "pop" only has a name and a return type ("Object").

For our example, consider executing the query and retrieving a result with the signature of Fig. 6.12.

This object has two method vectors $\vec{m}_{pushObject}$ and $\vec{m}_{popObject}$. In this case, the matching algorithm of Fig. 6.10 would match $\vec{m}_{push}$ to $\vec{m}_{pushObject}$ and $\vec{m}_{pop}$ to $\vec{m}_{popObject}$. Given the string preprocessing steps of Fig. 6.11 (e.g., "pushObject" would be split to "push" and "object", all names would become lowercase, etc.) and

```
public class MyStack {
    public bool pushObject(Object o){}
    public Object popObject(){}
}
```

Fig. 6.12 Example signature for a "Stack" component with two methods for adding and deleting elements from the stack

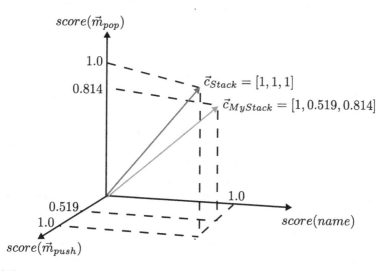

Fig. 6.13 Example vector space representation for the query "Stack" and the retrieved result "MyStack"

using the Levenshtein similarity for the strings, the method vectors $\vec{m}_{pushObject}$ and $\vec{m}_{popObject}$ would be

$$\vec{m}_{pushObject} = [0.429, 0, 1] \text{ and } \vec{m}_{popObject} = [0.429, 1] \qquad (6.9)$$

respectively. Using (6.5), the scores for $\vec{m}_{pushObject}$ and $\vec{m}_{popObject}$ would be 0.519 and 0.814, respectively. Hence, the class vector for $\vec{c}_{MyStack}$ would be

$$\vec{c}_{MyStack} = [1, 0.519, 0.814] \qquad (6.10)$$

Finally, given Eq. 6.5, the vectors of Eqs. 6.6 and 6.10 would provide score equal to 0.898 for the retrieved component. Figure 6.13 graphically illustrates how the class vectors are represented in the three-dimensional space for this example.

6.3.4.3   Postprocessor

Upon having scored the results, the postprocessor can be used to further enhance the produced ranking. Intuitively, when one or more files are functionally equivalent, a developer would probably select the simplest solution. As a result, we use the physical lines of code metric,[11] counting any line that has more than three non-blank characters, to determine the size and complexity of each file. After that, the results are ranked according to their functional score as computed by the scorer, in descending order, and subsequently in ascending order of their LoC value when the scores are equal.

6.3.5   Transformer

The transformer receives as input the retrieved software components and the query of the developer and performs a series of transformations so that the query is matched (i.e., the retrieved components are compatible along with the query). The transformations are performed on the srcML XML tree using XPath expressions and algorithms, and the resulting transformed XML produces the new source code component.

At first, class and method names are renamed to match those of the query. The method invocations are also examined to support cases where one method of the component calls another method that has to be renamed. Furthermore, the return types of methods are converted to the corresponding ones of the query, using type casts[12] wherever possible, while return statements are also transformed accordingly.

Upon having transformed the names and the return types of each method, the next step involves transforming the parameters. The parameters of the results have already been matched to those of the query using the algorithm of Fig. 6.10. Thus, the transformer iterates over all parameters of the query method and refactors the corresponding parameters of the result method. The refactoring involves changing the name, as well as the type of each parameter using type casts. Any parameters that are not present in the result method are added, while any extraneous parameters are removed from the parameter list and placed inside the method scope as variable declaration commands and are initialized to the default values as defined by Java standards.[13] Note also that the order of the parameters is changed to match the parameter order of the corresponding query method. These changes are also performed on all commands including invocations of the method.

---

[11] https://java.net/projects/loc-counter/pages/Home.

[12] http://imagejdocu.tudor.lu/doku.php?id=howto:java:how_to_convert_data_type_x_into_type_y_in_java.

[13] https://docs.oracle.com/javase/tutorial/java/nutsandbolts/datatypes.html.

```
public class Customer {
   public void setAddress(String i){}
   public String getAddress(){}
}
```

Fig. 6.14   Example signature for a class "Customer" with methods "setAddress" and "getAddress"

```
public class Customer {
   private String address;
   public Customer(){}
   public boolean setTheAddress(String address, String postCode) {
      this.address = address;
      if (postCode != null) {
         this.address += ", " + postCode;
         return true;
      }
      return false;
   }
   public String getTheAddress() {
      return address;
   }
}
```

Fig. 6.15   Example retrieved result for component "Customer"

Finally, using srcML, the transformed XML tree is parsed in order to extract the new version of the component. The code is further style-formatted using the Artistic Style formatter[14] to produce a clean ready-to-use component for the developer.

Figure 6.14 provides an example of a query signature for a "Customer" component with two methods, "setAddress" and "getAddress", that are used to set and retrieve the address of the customer, respectively. This example is used throughout this subsection to illustrate the functionality of the transformer module.

Figure 6.15 depicts an example retrieved result for the "Customer" component. This result indeed contains the required methods, however with slightly different names, different parameters, and different return types.

In this case, the transformer initially renames the methods setTheAddress and getTheAddress to setAddress and getAddress, respectively. After that, the return type of setAddress is changed from int to void and the corresponding return statements are modified. Finally, the extraneous parameter postCode is moved inside

---

[14]http://astyle.sourceforge.net/.

Fig. 6.16 Example modified
result for the file of Fig. 6.15

```
public class Customer {
    private String address;
    public Customer(){}
    public void setAddress(String address) {
        String postCode = null;
        this.address = address;
        if (postCode != null) {
            this.address += ", " + postCode;
            return;
        }
        return;
    }
    public String getAddress() {
        return address;
    }
}
```

the method and set to its default value (in this case null). The final result is shown in
Fig. 6.16.

6.3.6 Tester

The tester module of Mantissa evaluates whether the results satisfy the functional
conditions set by the developer. In order for this module to operate, the developer
has to provide test cases for the software component of the query.

Initially, a project folder is created for each result file. The corresponding result
file and the test case written by the developer are copied to the folder. After that, the
tester component employs the heuristics defined by the transformer in Sect. 6.3.5 to
modify the source code file (transform classes, methods, and parameters), so that the
file is compilable along with the test file. The files are compiled (if this is possible)
and the test is executed to provide the final result. The outcome of this procedure
indicates whether the downloaded result file is compilable and whether it passes the
provided test cases.

The corresponding test file that is provided by the developer for the component
of Fig. 6.14 is shown in Fig. 6.17.

The test is formed using JUnit and the Java *reflection* feature. Reflection is required
to provide a test structure that can be modified in order to comply with different files.
In the code of Fig. 6.17, variable m_index is a zero-based counter, which stores the
position of the method that is going to be invoked next in the result file. Using a

```
import static org.junit.Assert.assertEquals;
import org.junit.Test;
import java.lang.reflect.*;

public class CustomerTest {
    @Test
    public void testAddress() throws Exception {
        int m_index = 0;
        Class<?> clazz = Class.forName("Customer");
        Object c = clazz.newInstance();
        Method m = clazz.getDeclaredMethod("setAddress", String.class);
        m.invoke(c,"test");
        m_index = 1;
        m = clazz.getDeclaredMethod("getAddress");
        assertEquals("Wrong string!","test",m.invoke(c));
    }
}
```

Fig. 6.17 Example test case for the "Customer" component of Fig. 6.14

counter eliminates any issues with overloaded methods.[15] An object is created by retrieving one of its constructors using the getDeclaredConstructor function and invoking it by calling the newInstance function. After that, methods are retrieved using the getDeclaredMethod function, including any parameter types, and finally they are invoked using the invoke function.

Before compiling any new file along with the test case, the tester has to modify the files in order to be compliant. The modification of results and test files includes renaming the packages and transforming all access modifiers from private/protected to public in order to allow compiling the retrieved file along with the test file. The retrieved file needs no further transformations since the transformer has already performed all refactorings on class, method, and parameter levels.

The next step is to compile the files. The Eclipse compiler[16] is used since it allows compiling multiple files, without breaking in case of errors in a single file. Finally, for the files that are successfully compiled, the test cases are executed. The final results including their status, whether they are compilable and whether they passed the tests, are provided to the developer.

---

[15] Java supports method overloading, meaning that it allows the existence of methods with the same name that differs only in their parameters.

[16] http://www.eclipse.org/jdt/core/.

6.3.7 Flow Extractor

The flow extractor extracts the source code flow of the methods for each result. The types that are used in each method and the sequence of commands can be a helpful asset for the developer, as he/she may easily understand the inner functionality of the method, without having to read the source code. We use the sequence extractor of [45], and further extend it to account for conditions, loops, and try-catch statements.

In specific, for each component, we initially extract all the declarations from the source code (including classes, fields, methods, and variables) to create a hierarchical lookup table according to the scopes as defined by the Java language. After that, we iterate over the statements of each method and extract three types of statements: assignments, functions calls, and class instantiations. For instance, the command this.address = address of Fig. 6.16 is an assignment statement of type String. All different alternate paths are taken into account. Apart from conditions, try-catch statements define different flows (with finally blocks executed in all cases), while the loops are either executed or not, i.e., are regarded as conditions.

Finally, for each method, its possible flows are depicted in the form of a graph, where each node represents a statement. The graphs are constructed in the Graphviz dot format[17] and generated using Viz.js.[18] The types of the nodes are further printed in red font when they refer to external dependencies. An example usage of the flow extractor will be given in Sect. 6.4.2 (see Figs. 6.21 and 6.22).

6.4 Mantissa User Interface and Search Scenario

In this section, we present the user interface of Mantissa (Sect. 6.4.1) and provide an example of using our tool for a realistic search scenario (Sect. 6.4.2).

6.4.1 Mantissa User Interface

The prototype user interface of Mantissa is a client-server application. The server is created in Python using Flask,[19] while HTML and JavaScript have been used to develop the user interface of the service. The main page of Mantissa is shown in Fig. 6.18.

As shown in this figure, the search page provides two text areas in the center of the screen that allow the user to enter an interface query and the test code that will be used to evaluate the retrieved components. There is also the option of not providing a test file, which can be selected by activating the "No test case" checkbox. Finally, at

---

[17]http://www.graphviz.org/.

[18]http://viz-js.com/.

[19]http://flask.pocoo.org/.

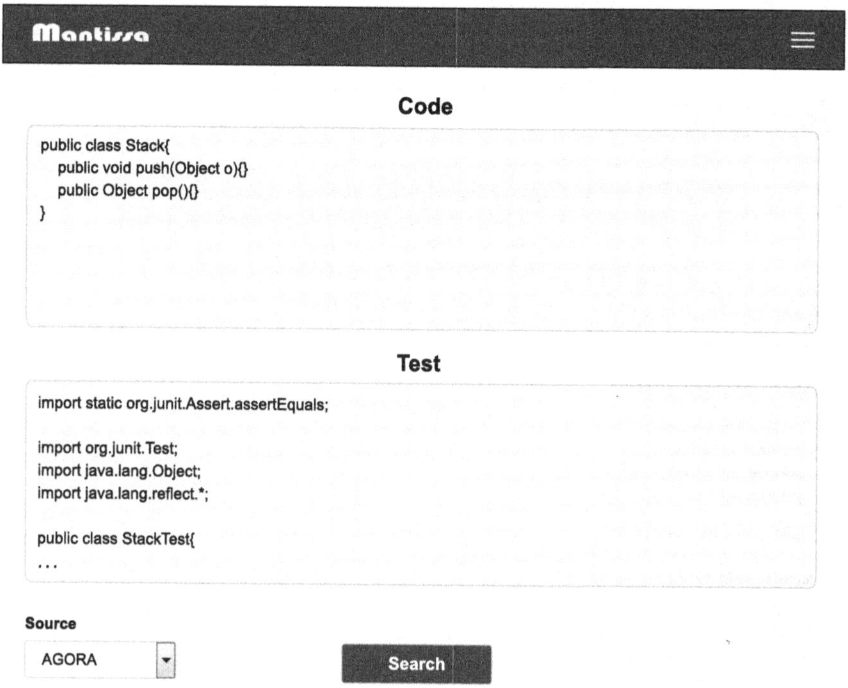

Fig. 6.18 Screenshot depicting the search page of Mantissa

the bottom of the screen, the user can select the CSE to be used for searching, either AGORA or GitHub, using a dropdown menu, before hitting the "Search" button. In the example of Fig. 6.18, the user has entered a query for a "Stack" component and the corresponding test case.

Upon issuing a search, relevant results are presented to the user in the results page of Fig. 6.19. As shown in this figure, the user can navigate the results as well as possibly integrate any useful component in his/her code. Each result is accompanied by information about the score given by the system, its physical lines of code, whether it compiled and whether it successfully passed the tests. For instance, Fig. 6.19 depicts the nineteenth result for the "Stack" query, which has score value equal to 0.96, consists of 27 lines of code, while it compiled but did not pass the test. Finally, the provided options (in the form of links) allow viewing the original file (i.e., before any transformation by the transformer), showing the control flow for each function of the file and showing the external dependencies in the form of imports.

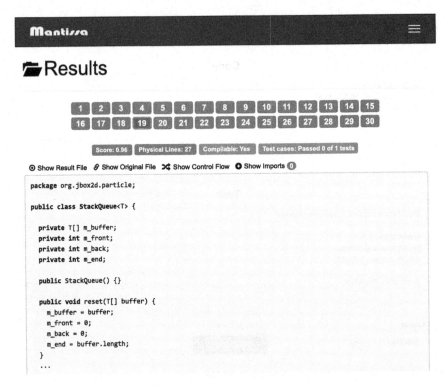

Fig. 6.19 Screenshot depicting the results page of Mantissa

6.4.2 Example Search Scenario

In this section, we provide an example usage scenario to illustrate how Mantissa can be a valuable asset for the developer. The scenario involves creating a duplicate file check application. The application will initially scan the files of a given folder and store the MD5 representation for the contents of each file in a set. After that, any new files will be checked using the set before being added to the folder.

The first step is to divide the application into different components. Typically, the developer would split the application in a file reader/writer module, a serializable set data structure that can be saved to/loaded from disk, and an implementation of the MD5 algorithm. Using Mantissa, we are able to search for and retrieve these components quite easily. For the file reader/writer, we construct the query shown in Fig. 6.20.

A quick check in the results list reveals that several components provide this functionality. In this case, the one selected would be the smaller one out of all components that are compilable. Indicatively, the file write method of the file tools component is shown in Fig. 6.21.

```
public class FileTools {
    public String read(String filename){}
    public void write(String filename, String content){}
}
```

Fig. 6.20 Example signature for a component that reads and writes files

```
public static void write(String content, String filePath) throws IOException {
    BufferedWriter out = null;
    try {
        out = new BufferedWriter(new FileWriter(filePath));
        out.write(content);
    } finally {
        if (out != null)
            out.close();
    }
}
```

Fig. 6.21 Example write method of a component that reads and writes files

Further examining the source code flow of the methods for the selected result indicates that methods of the Java standard library are employed, and therefore the test, in this case, can be omitted for simplicity. For example, Fig. 6.22 depicts the code flow for the method shown in Fig. 6.21, where it is clear that the main type involved is the BufferedWriter, which is included in the standard library of Java.

For the MD5 algorithm, the query is shown in Fig. 6.23, while the corresponding test case is shown in Fig. 6.24. The test involves checking the correct hashing of a known string. In this case, the second result passes the test case and has no external dependencies.

The next component is the serializable set data structure. The query in this case is shown in Fig. 6.25 and the corresponding test case is shown in Fig. 6.26. Note how the main functionality lies on the serialization part, and therefore this is covered by the test case.

Finally, upon having downloaded the components, the developer can now easily write the remaining code for creating the MD5 set, reading or writing it to disk, and checking new files for duplicate information. The code for creating the database is shown in Fig. 6.27.

Overall, the developer has to write less than 50 lines of code in order to construct both the queries and the tests for this application. Upon finding executable components that cover the desired functionality, the integration effort required is also minimal. Finally, the main flow of the application (a part of which is shown in Fig. 6.27) is quite simple, given that all implementation details of the components

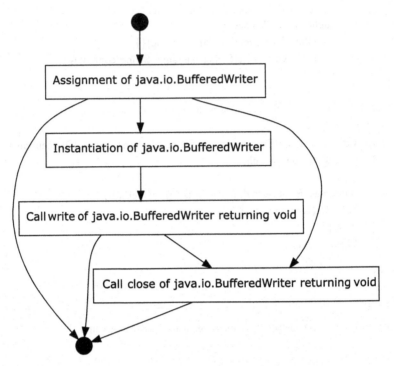

Fig. 6.22 Code flow for the write method of a component that handles files

Fig. 6.23 Example signature
for a component that finds
the MD5 of a string

```
public class Md5 {
        public String md5Sum(String i){}
}
```

are very well abstracted. Embracing this component-based development paradigm can result in reduced time and effort for the developer and overall cleaner design. Additionally, employing the test-driven characteristics of Mantissa, the developer can further be certain that the components cover the required functionality.

6.5 Evaluation

We evaluate Mantissa against popular CSEs (the ones assessed also in the previous chapter, i.e., GitHub Search, BlackDuck Open Hub, and AGORA), as well as against two well-known RSSEs (FAST and Code Conjurer). For each of the three experiments, we define and present the dataset used for the evaluation, while the results are summarized in tables and illustrated in figures. Note that each experiment has a

```
import static org.junit.Assert.assertEquals;
import org.junit.Test;
import java.lang.reflect.*;

public class Md5Test {

    @Test
    public void testMd5Sum() throws Exception {
        int m_index = 0;
        Class<?> c = Class.forName("Md5");
        Method m = c.getDeclaredMethod("md5Sum", String.class);
        assertEquals("c2a9ce57e8df081b4baad80d81868bbb",
                m.invoke(null, "This is my string"));
    }
}
```

Fig. 6.24 Example test case for the "MD5" component of Fig. 6.23

Fig. 6.25 Example signature for a serializable set component

```
public class SerializableSet {
    public String serialize(){}
    public SerializableSet deserialize(String s){}
}
```

different dataset since some of the systems are no longer available, thus only their own datasets could be used for evaluating them.

6.5.1 Evaluation Framework

Our evaluation is based on the scenario of searching for a reusable software component, as this was described in the evaluation of the previous chapter. In our scenario, the developer provides a CSE or an RSSE with a query for the desired component. The system returns relevant results and the developer examines them and determines whether they are relevant to the query. As before, we refrain from examining the results, since this would pose serious threats to validity, and instead determine whether a result is relevant by using automated testing (like with the tester module of Mantissa). As a result, we do not assess the tester. The datasets used for evaluation consist of queries for components and their corresponding test cases.

For each query, we use the number of *compiled* results and the number of results that *passed the tests* as metrics. Furthermore, as in the previous chapter, we employ

```java
import static org.junit.Assert.assertEquals;
import org.junit.Test;
import java.lang.reflect.*;

public class SerializableSetTest {

    @Test
    public void testSerialization() throws Exception {
        int m_index = 0;
        Class<?> clazz = Class.forName("SerializableSet");
        Object c = clazz.newInstance();
        Method m = clazz.getDeclaredMethod("serialize");
        Object serializedSet = m.invoke(c);
        m_index = 1;
        m = clazz.getDeclaredMethod("deserialize", String.class);
        Object deserializedSet = m.invoke(c, serializedSet);
        assertEquals(c, deserializedSet);
    }
}
```

Fig. 6.26 Example test case for the serializable set component of Fig. 6.25

```java
SerializableSet set = new SerializableSet();
for (File string : new File(folderPath).listFiles()) {
    String content = FileTools.read(string.getAbsolutePath());
    String md5String = Md5.md5Sum(content);
    set.add(md5String);
}
FileTools.write(set.serialize(), setFilename);
```

Fig. 6.27 Example for constructing a set of MD5 strings for the files of folder folderPath and writing it to a file with filename setFilename

the search length, which we define as the number of non-compilable results (or results that did not pass the tests) that the developer must examine in order to find a number of compilable results (or results that passed the tests). We again assess all the metrics for the first 30 results for each system, since keeping more results did not have any impact on the compilability or the test passing score, while 30 results are usually more than a developer would examine in such a scenario.

6.5.2 Comparing Mantissa with CSEs

In this section, we extend our evaluation of CSEs presented in the previous chapter in order to illustrate how using Mantissa improves on using only a CSE. As such, we compare Mantissa against the well-known CSEs of GitHub and BlackDuck as well as against our CSE, AGORA.

6.5.2.1 Dataset

Our dataset consists of 16 components and their corresponding test cases. These components are of variable complexity to fully assess the results of the CSEs against Mantissa in the component reuse scenario. Our dataset is shown in Table 6.3.

The formation of the queries for the three CSEs is identical to the one described in the previous chapter. The dataset, however, is updated to include 16 components. The original 13 of them refer to classes, while the 3 new ones, "Prime", "Sort", and "Tokenizer", refer to functions. These components were added to further assess the CSEs against Mantissa in scenarios where the required component is a function and not a class. Further analyzing the dataset, we perceive that there are simple queries, some of them related to mathematics, for which we anticipate numerous results

Table 6.3 Dataset for the evaluation of Mantissa against code search engines

Class	Methods
Account	deposit, withdraw, getBalance
Article	setId, getId, setName, getName, setPrice, getPrice
Calculator	add, subtract, divide, multiply
ComplexNumber	ComplexNumber, add, getRealPart, getImaginaryPart
CardValidator	isValid
Customer	setAddress, getAddress
Gcd	gcd
Md5	md5Sum
Mortgage	setRate, setPrincipal, setYears, getMonthlyPayment
Movie	Movie, getTitle
Prime	checkPrime
Sort	sort
Spreadsheet	put, get
Stack	push, pop
Stack2	pushObject, popObject
Tokenizer	tokenize

(e.g., "Gcd" or "Prime"), and more complex queries (e.g., "Mortgage") or queries with rare names (e.g., "Spreadsheet") that are more challenging for the systems. Finally, although some queries might seem simple (e.g., "Calculator"), they may have dependencies from third-party libraries that would lead to compilation errors.

6.5.2.2   Results

The results of the comparison of Mantissa with the CSEs are shown in Table 6.4. At first, one may notice that both GitHub and AGORA, as well as the two implementations of Mantissa, returned at least 30 results for each query. By contrast, BlackDuck returned fewer results in half of the queries, indicating that its retrieval mechanism performs exact matching, and therefore is more sensitive to small changes in the query. This is particularly obvious for the modified version of the "Stack" query, where the service returned no results.

GitHub seems to return several results, however few of them are compilable, while most queries return results that do not pass the tests. This is expected since the

Table 6.4 Compiled and passed results out of the total returned results for the three code search engines and the two Mantissa implementations, for each result and on average, where the format is passed/compiled/total results

Query	Code search engines			Mantissa	
	GitHub	AGORA	BlackDuck	AGORA	GitHub
Account	0/10/30	1/4/30	2/9/30	1/5/30	6/7/30
Article	1/6/30	0/5/30	2/3/15	5/9/30	2/8/30
Calculator	0/5/30	0/5/30	1/3/22	0/4/30	0/9/30
ComplexNumber	2/2/30	2/12/30	1/1/3	4/7/30	7/11/30
CardValidator	0/1/30	1/4/30	2/3/30	0/0/30	0/4/30
Customer	2/3/30	13/15/30	7/7/30	20/21/30	5/6/30
Gcd	1/3/30	5/9/30	12/16/30	5/9/30	20/22/30
Md5	3/7/30	2/10/30	0/0/0	4/7/30	1/3/30
Mortgage	0/6/30	0/4/30	0/0/0	0/7/30	0/1/30
Movie	1/3/30	2/2/30	3/7/30	2/2/30	1/5/30
Prime	4/13/30	1/4/30	1/2/15	1/1/30	4/14/30
Sort	0/7/30	0/1/30	0/4/30	1/4/30	4/5/30
Spreadsheet	0/1/30	1/2/30	0/0/6	1/2/30	1/3/30
Stack	0/3/30	7/8/30	8/12/30	14/15/30	10/11/30
Stack2	0/0/30	7/8/30	0/0/0	14/15/30	1/1/30
Tokenizer	0/8/30	0/4/30	2/7/30	0/3/30	1/4/30
Average results	30.0	30.0	18.8	30.0	30.0
Average compiled	4.875	6.062	4.625	6.938	7.125
Average passed	0.875	2.625	2.562	4.500	3.938

Fig. 6.28 Evaluation diagram depicting the average number of compilable and tested results for the three code search engines and the two Mantissa implementations

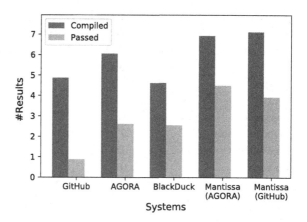

service is not optimized for component-retrieval scenarios. AGORA, on the other hand, succeeds in producing relevant and tested results for most queries. The service retrieves approximately six compilable results per query, while it is also the only CSE to produce at least one compilable result for every query, even the modified "Stack" one.

As we can see, integrating AGORA to our system almost doubled the number of results that pass the tests, while integrating GitHub quadrupled the aforementioned metric. Both implementations of Mantissa are quite effective, retrieving at least one compilable component for almost all queries, and at least one that passes the tests for 13 out of 16 queries.

Further analyzing the results, we see that both implementations of Mantissa returned numerous results for single method queries (e.g., "Gcd" and "Sort") or generally simple queries (e.g., "Customer" and "Stack"). The GitHub implementation seems superior in mathematical queries, such as "Gcd" and "Prime", which is expected since the index of AGORA is much smaller than that of GitHub. Furthermore, mathematical components may also have dependencies from third-party libraries that would result in compilation errors, which is, however, noticeable in both services, as in the "Calculator" component. Note, though, that Mantissa also provides links to grepcode for the identified dependencies, which could be helpful for such queries. The implementation using AGORA is more effective for challenging queries, such as the "Stack2" query, which is due to the lexical analysis performed by the service.

Finally, the average number of the compilable results and results that passed the tests for each of the five systems are visualized in Fig. 6.28. From this figure, it is also clear that the implementations of Mantissa are more effective than the three CSEs.

Apart from the performance of the five systems on retrieving useful results, it is also important to determine whether these results are ranked in an effective order. Given that a tested result is probably the most useful for the developer, the search length is computed with regard to finding the first result that passed the tests. The results for the search length are shown in Table 6.5 for the five systems.

Table 6.5 Search length for the passed results for the three code search engines and the two Mantissa implementations, for each result and on average

Query	Code search engines			Mantissa	
	GitHub	AGORA	BlackDuck	AGORA	GitHub
Account	30	2	0	0	0
Article	21	30	2	11	1
Calculator	30	30	0	30	30
ComplexNumber	0	0	0	0	0
CardValidator	30	15	2	30	30
Customer	0	0	1	0	0
Gcd	25	8	5	0	0
Md5	9	0	30	1	2
Mortgage	30	30	30	30	30
Movie	23	21	1	22	7
Prime	1	10	0	6	0
Sort	30	30	30	0	1
Spreadsheet	30	23	30	8	9
Stack	30	2	4	0	3
Stack2	30	2	30	0	0
Tokenizer	30	30	14	30	8
Average passed	21.812	14.562	11.188	10.500	7.562

Both Mantissa implementations placed a successfully tested result in the first position in almost half of the dataset queries. Even in queries where only a single result passed the tests, such as "Prime" and "Sort", this was ranked in a high position. Both implementations are more effective than any CSE. Concerning the three CSEs, GitHub falls behind the other systems, while AGORA and BlackDuck seem to provide effective rankings in roughly half of the queries.

The average search length for finding more than one result that passed the tests is shown in Fig. 6.29. As shown in this figure, BlackDuck is effective for finding the first relevant result, while AGORA is better when more results are required. GitHub is outperformed by all systems. Comparing all systems, it is clear that using Mantissa requires examining fewer irrelevant results to find the first relevant ones. Given, for example, that a developer wants to find the first three results that passed the tests, then on average he/she would have to examine 18.6 of the results of Mantissa using GitHub and 18.9 using AGORA in contrast with the CSEs where he/she would have to examine 23.5 of the results of BlackDuck, 24.4 of the results of AGORA, and 29.13 of the results of GitHub. The Mantissa implementation using GitHub outperforms all other systems for finding the first results, while the AGORA implementation is more effective for finding slightly more relevant results.

Fig. 6.29 Evaluation diagram depicting the average search length for finding results that pass the tests for the three code search engines and the two Mantissa implementations

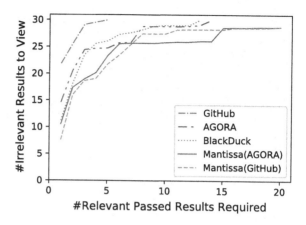

6.5.3 Comparing Mantissa with FAST

In this section, we evaluate Mantissa against the FAST plugin [34]. We use the dataset provided for the evaluation of FAST for this comparison since the plugin employs the deprecated CSE Merobase so it is not functional.

6.5.3.1 Dataset

The dataset of FAST consists of 15 different components and 15 corresponding test cases. The components are shown in Table 6.6, where it is clear that most queries contain only a single method and are mostly related to mathematical tasks.

Queries with more than one method, such as the "FinancialCalculator" class, are quite complex, so they are not expected to have adequate results. An interesting query is "StackInit" which differs from "Stack" in that it contains a constructor. Moreover, the queries "Sort" and "SumArray" may return classes whose methods differ in parameters type (e.g., List<Integer> instead of int[]). Since FAST is not functional, we were not able to compute the search length metric. Consequently, we evaluate the two systems based on the number of results that passed compilations and the number of results that passed the tests.

6.5.3.2 Results

Table 6.7 presents the results for Mantissa and FAST, where it is clear that Mantissa returns more compilable results as well as more results that pass the tests than the FAST system. Further analyzing the results, FAST is effective for only half of the queries, while there are several queries where its results are not compilable. Both implementations returned numerous tested results, even in cases where FAST was not

Table 6.6 Dataset for the evaluation of Mantissa against FAST

Class	Methods
CardValidator	isValid
Customer	setName, getName, setId, getId, setBirthday, getBirthday
Fibonacci	Fibonacci
FinancialCalculator	setPaymentsPerYear, setNumPayments, setInterest, setPresentValue, setFutureValue, getPayment
Gcd	gcd
isDigit	isDigit
isLeapYear	isLeapYear
isPalindrome	isPalindrome
Md5	md5Sum
Mortgage	calculateMortgage
Prime	isPrime
Sort	sort
Stack	push, pop
StackInit	Stack, push, pop
SumArray	sumArray

effective (e.g., "Md5" and "SumArray" classes). Similar to our previous experiment, mathematical queries, such as "FinancialCalculator" and "Prime", were challenging for all systems, possibly indicating dependencies from third-party libraries. It is worth noticing that the GitHub implementation of Mantissa seems more promising than the AGORA one. However, this is expected since GitHub has millions of repositories, while the index of AGORA contains fewer repositories that may not contain mathematical functions, as in the dataset of FAST.

The average number of compilable results and passed results are also shown in Fig. 6.30, where it is clear that the Mantissa implementations outperform FAST.

6.5.4 Comparing Mantissa with Code Conjurer

In this subsection, we evaluate Mantissa against Code Conjurer [13]. Code Conjurer has two search approaches: the "Interface-based" approach and the "Adaptation" approach. The first conducts exact matching between the query and the retrieved components, while the second conducts flexible matching, ignoring method names, types, etc. Similarly to FAST, Code Conjurer is also currently not functional, and thus we use the provided dataset to compare it to our system. Finally, since the number of compilable results or the ranking of the results is not provided for Code Conjurer, we compare the systems only with respect to the number of results that pass the tests.

Table 6.7 Compiled and passed results out of the total returned results for FAST and the two Mantissa implementations, for each result and on average, where the format for each value is passed/compiled/total results

Query	FAST	Mantissa	
		AGORA	GitHub
CardValidator	1/1/1	0/0/30	0/4/30
Customer	0/0/5	10/10/30	0/3/30
Fibonacci	4/11/30	2/2/30	4/12/30
FinancialCalculator	0/0/0	0/5/30	0/2/30
Gcd	12/12/30	5/9/30	20/22/30
isDigit	2/2/30	6/6/30	10/12/30
isLeapYear	4/4/30	1/2/30	2/3/30
isPalindrome	4/4/9	1/3/30	6/18/30
Md5	0/0/24	4/7/30	1/3/30
Mortgage	0/0/0	0/2/30	0/5/30
Prime	3/4/30	1/1/30	5/14/30
Sort	1/1/30	1/4/30	4/5/30
Stack	10/10/30	12/14/30	9/14/30
StackInit	3/4/30	5/12 /30	5/14/30
SumArray	0/0/30	5/9/30	1/8/30
Average results	20.6	30.0	30.0
Average compiled	3.533	5.733	9.267
Average passed	2.933	3.533	4.467

Fig. 6.30 Evaluation diagram depicting the average number of compilable and tested results for FAST and for the two Mantissa implementations

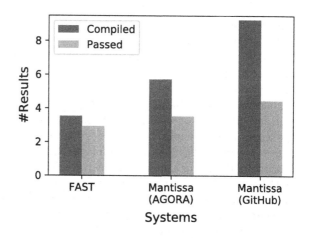

Table 6.8 Dataset for the evaluation of Mantissa against Code Conjurer

Class	Methods
Calculator	add, sub, div, mult
ComplexNumber	ComplexNumber, add, getRealPart, getImaginaryPart
Matrix	Matrix, set, get, multiply
MortgageCalculator	setRate, setPrincipal, setYears, getMonthlyPayment
Spreadsheet	put, get
Stack	push, pop

6.5.4.1   Dataset

The dataset, which is shown in Table 6.8, consists of six different components and six corresponding test cases. Compared to the dataset of [13], we omitted the "Shopping-Cart" component because it depends on the external "Product" class, and therefore it could not be executed. The dataset is rather limited; however, it includes different types of queries that could lead to interesting conclusions. Complex queries with multiple methods, such as "ComplexNumber" or "MortgageCalculator", are expected to be challenging for the systems. On the other hand, queries that include mathematical components, such as "Calculator" or "Matrix", may involve dependencies from third-party libraries, and hence they may also be challenging for the systems.

6.5.4.2   Results

The results of our comparison, shown in Table 6.9, are highly dependent on the given queries for all systems. One of the main advantages of Code Conjurer, especially of its adaptation approach, is its ability to find numerous results. The approach retrieves the maximum number of files that comply with the given query from the Merobase CSE and attempts to execute the test cases by performing certain transformations. In this scope, it is not comparable with our system since Mantissa uses certain thresholds as to the number of the files retrieved from the CSEs. This allows Mantissa to present meaningful results for a query in a timely manner, i.e., at most 3 min.

Using the maximum number of files has a strong impact on the response time of Code Conjurer. The adaptation approach in some cases requires several hours to return tested components. The interface-based approach provides results in a more timely manner, including, however, unexpected deviations in the response time among different queries (e.g., 26 min for the "Stack" query).

Concerning the effectiveness of the interface-based approach of Code Conjurer in comparison with Mantissa, it is worth noticing that each system performs better in different queries. Code Conjurer is more effective for finding tested components

Table 6.9 Results that passed the tests and response time for the two approaches of Code Conjurer (Interface-based and Adaptation) and the two implementations of Mantissa (using AGORA and using GitHub), for each result and on average

Query	Code Conjurer				Mantissa			
	Interface		Adaptation		AGORA		GitHub	
	Passed	Time	Passed	Time	Passed	Time	Passed	Time
Calculator	1	19 s	22	20 h 24 m	0	25 s	0	2 m 35 s
ComplexNumber	0	3 s	30	1 m 19 s	4	35 s	7	2 m 8 s
Matrix	2	23 s	26	5 m 25 s	0	31 s	0	2 m 51 s
MortgageCalculator	0	4 s	15	3 h 19 m	0	31 s	0	2 m 18 s
Spreadsheet	0	3 s	4	15 h 13 m	1	23 s	1	2 m 28 s
Stack	30	26 m	30	18 h 23 m	14	30 s	10	2 m 53s
Average	5.500	4 m 28 s	21.167	9 h 34 m 17 s	3.167	29 s	3.000	2 m 32 s

for the "Calculator" and "Matrix" queries, while both implementations of Mantissa retrieve useful components for the "ComplexNumber" and "Spreadsheet" queries. Notably, all systems found several results for the "Stack" query, while "Mortgage-Calculator" was too complex to produce tested components in any system. Note, however, that Code Conjurer retrieved and tested 692 components for this query (requiring 26 min), whereas the Mantissa implementations were restricted to 200 components.

Finally, given that the dataset contains only six queries, and given that using a different dataset was not possible since Merobase is not functional, it is hard to arrive at any conclusions about the effectiveness of the systems. However, this experiment indicates that Mantissa has a steady response time. As shown in Table 6.9, the response time of the AGORA implementation is approximately 35 s, while the GitHub implementation returns tested components within roughly 2.5 min. Both implementations are consistent, having quite small deviations from these averages.

6.6 Conclusion

In this chapter, we designed and implemented Mantissa, an RSSE that extracts the query from developer code and recommends reusable components. The mining model of our system is combined with a transformer that matches the components to the query of the developer, while a robust tester module is employed to test the functionality of the retrieved source code files. Our evaluation showed that Mantissa outperforms CSEs, while its comparison with two RSSEs indicates that it presents useful results in a timely manner and outperforms the other systems in several scenarios.

Future work on Mantissa lies in several directions. Regarding the tester component, it would be interesting to work on resolving component dependencies from libraries in order to increase the number of results that could be compiled and pass the tests. Additionally, hosting a developer survey could lead to better assessing Mantissa and understanding which modules are of most value to the developers.

References

1. Walker RJ (2013) Recent advances in recommendation systems for software engineering. In: Ali M, Bosse T, Hindriks KV, Hoogendoorn M, Jonker CM, Treur J (eds) Recent trends in applied artificial intelligence, vol 7906. Lecture notes in computer science. Springer, Berlin, pp 372–381
2. Nurolahzade M, Walker RJ, Maurer F (2013) An assessment of test-driven reuse: promises and pitfalls. In: Favaro John, Morisio Maurizio (eds) Safe and secure software reuse, vol 7925. Lecture notes in computer science. Springer, Berlin Heidelberg, pp 65–80
3. McIlroy MD (1968) Components mass-produced software. In: Naur P, Randell B (eds.) Software engineering; report of a conference sponsored by the nato science committee, pp 138–155. NATO Scientific Affairs Division, Brussels, Belgium, NATO Scientific Affairs Division. Belgium, Brussels
4. Mens K, Lozano A (2014) Source code-based recommendation systems. Springer, Berlin, pp 93–130
5. Janjic W, Hummel O, Atkinson C (2014) Reuse-oriented code recommendation systems. Springer, Berlin, pp 359–386
6. Robillard M, Walker R, Zimmermann T (2010) Recommendation systems for software engineering. IEEE Softw 27(4):80–86
7. Gasparic M, Janes A (2016) What recommendation systems for software engineering recommend. J Syst Softw 113(C):101–113
8. Sahavechaphan N, Claypool K (2006) XSnippet: mining for sample code. SIGPLAN Not 41(10):413–430
9. Thummalapenta S, Xie T (2007) PARSEWeb: a programmer assistant for reusing open source code on the web. In: Proceedings of the 22nd IEEE/ACM international conference on automated software engineering, ASE '07, pp. 204–213, New York, NY, USA. ACM
10. Xie T, Pei J (2006) MAPO: mining API usages from open source repositories. In: Proceedings of the 2006 international workshop on mining software repositories, MSR '06, pp 54–57, New York, NY, USA. ACM
11. Wei Y, Chandrasekaran N, Gulwani S, Hamadi Y (2015) Building bing developer assistant. Technical Report MSR-TR-2015-36, Microsoft Research
12. Galenson J, Reames P, Bodik R, Hartmann B, Sen K (2014) CodeHint: dynamic and interactive synthesis of code snippets. In: Proceedings of the 36th international conference on software engineering, ICSE 2014, pp 653–663, New York, NY, USA. ACM
13. Hummel O, Janjic W, Atkinson C (2008) Code conjurer: pulling reusable software out of thin air. IEEE Softw 25(5):45–52
14. Lemos OAL, Bajracharya SK, Ossher J, Morla RS, Masiero PC, Baldi P, Lopes CV (2007) CodeGenie: using test-cases to search and reuse source code. In: Proceedings of the Twenty-second IEEE/ACM international conference on automated software engineering, ASE '07, pp 525–526, New York, NY, USA. ACM

15. Henninger S (1991) Retrieving software objects in an example-based programming environment. In: Proceedings of the 14th Annual International ACM SIGIR Conference on Research and Development in Information Retrieval, SIGIR '91, pp 251–260, New York, NY, USA. ACM

16. Page L, Brin S, Motwani R, Winograd T (1999) The PageRank citation ranking: bringing order to the web. Technical Report 1999-66, Stanford InfoLab. Previous number = SIDL-WP-1999-0120

17. Michail A (2000) Data mining library reuse patterns using generalized association rules. In: Proceedings of the 22nd international conference on software engineering, ICSE '00, pp 167–176, New York, NY, USA. ACM

18. Ye Y, Fischer G (2002) Supporting reuse by delivering task-relevant and personalized information. In: Proceedings of the 24th international conference on software engineering, ICSE '02, pp 513–523, New York, NY, USA. ACM

19. Holmes R, Murphy GC (2005) Using structural context to recommend source code examples. In: Proceedings of the 27th international conference on software engineering, ICSE '05, pp 117–125, New York, NY, USA. ACM

20. Mandelin D, Lin X, Bodík R, Kimelman D (2005) Jungloid mining: helping to navigate the PI jungle. SIGPLAN Not 40(6):48–61

21. McMillan C, Grechanik M, Poshyvanyk D, Xie Q, Fu C (2011) Portfolio: finding relevant functions and their usage. In: Proceedings of the 33rd international conference on software engineering, ICSE '11, pp 111–120, New York, NY, USA. ACM

22. McMillan C, Grechanik M, Poshyvanyk D, Chen F, Xie Q (2012) Exemplar: a source code search engine for finding highly relevant applications. IEEE Trans Softw Eng 38(5):1069–1087

23. Wightman D, Ye Z, Brandt J, Vertegaal R (2012) SnipMatch: using source code context to enhance snippet retrieval and parameterization. In: Proceedings of the 25th annual ACM symposium on user interface software and technology, UIST '12, pp 219–228, New York, NY, USA. ACM

24. Zagalsky A, Barzilay O, Yehudai A (2012) Example overflow: using social media for code recommendation. In: Proceedings of the third international workshop on recommendation systems for software engineering, RSSE '12, pp 38–42, Piscataway, NJ, USA. IEEE Press

25. Beck (2002) Test driven development: by example. Addison-Wesley Longman Publishing Co., Inc., Boston

26. Hummel O, Atkinson C (2004) Extreme Harvesting: test driven discovery and reuse of software components. In: Proceedings of the 2004 IEEE international conference on information reuse and integration, IRI 2004, pp 66–72

27. Janjic W, Stoll D, Bostan P, Atkinson C (2009) Lowering the barrier to reuse through test-driven search. In: Proceedings of the 2009 ICSE workshop on search-driven development-users, infrastructure, tools and evaluation, SUITE '09, pp 21–24, Washington, DC, USA. IEEE Computer Society

28. Hummel O, Janjic W (2013) Test-driven reuse: key to improving precision of search engines for software reuse. In: Sim SE, Gallardo-Valencia RE (eds) Finding source code on the web for remix and reuse, pp 227–250. Springer, New York

29. Janjic W, Hummel O, Schumacher M, Atkinson (2013) An unabridged source code dataset for research in software reuse. In: Proceedings of the 10th working conference on mining software repositories, MSR '13, pp 339–342, Piscataway, NJ, USA. IEEE Press

30. Lemos OAL, Bajracharya S, Ossher J, Masiero PC, Lopes C (2009) Applying test-driven code search to the reuse of auxiliary functionality. In: Proceedings of the 2009 ACM symposium on applied computing, SAC '09, pp 476–482, New York, NY, USA. ACM

31. Lemos OAL, Bajracharya S, Ossher J, Masiero PC, Lopes C (2011) A test-driven approach to code search and its application to the reuse of auxiliary functionality. Inf Softw Technol 53(4):294–306

32. Bajracharya S, Ngo T, Linstead E, Dou Y, Rigor P, Baldi P, Lopes C (2006) Sourcerer: a search engine for open source code supporting structure-based search. In: Companion to the 21st ACM SIGPLAN symposium on object-oriented programming systems, languages, and applications, OOPSLA '06, pp 681–682, New York, NY, USA. ACM

33. Linstead E, Bajracharya S, Ngo T, Rigor P, Lopes C, Baldi P (2009) Sourcerer: mining and searching internet-scale software repositories. Data Min Knowl Discov 18(2):300–336
34. Krug M (2007) FAST: an eclipse plug-in for test-driven reuse. Master's thesis, University of Mannheim
35. Reiss SP (2009) Semantics-based code search. In: Proceedings of the 31st international conference on software engineering, ICSE '09, pp 243–253, Washington, DC, USA. IEEE Computer Society
36. Reiss SP (2009) Specifying what to search for. In: Proceedings of the 2009 ICSE workshop on search-driven development-users, infrastructure, tools and evaluation, SUITE '09, pp 41–44, Washington, DC, USA. IEEE Computer Society
37. Diamantopoulos T, Katirtzis N, Symeonidis A (2018) Mantissa: a recommendation system for test-driven code reuse. Unpublished manuscript
38. Rivest R (1992) The MD5 message-digest algorithm
39. Manning CD, Raghavan P, Schütze H (2008) Introduction to information retrieval. Cambridge University Press, New York
40. Gale D, Shapley LS (1962) College admissions and the stability of marriage. Am Math Mon 69(1):9–15
41. Bird S, Klein E, Loper E (2009) Natural language processing with python, 1st edn. O'Reilly Media, Inc., Sebastopol
42. Levenshtein VI (1966) Binary codes capable of correcting deletions, insertions and reversals. Soviet Physics Doklady 10:707
43. Jaccard P (1901) Étude comparative de la distribution florale dans une portion des alpes et des jura. Bulletin del la Société Vaudoise des Sciences Naturelles 37:547–579
44. Tanimoto TT (1957) IBM Internal Report
45. Diamantopoulos T, Symeonidis AL (2015) Employing source code information to improve question-answering in stack overflow. In: Proceedings of the 12th working conference on mining software repositories, MSR '15, pp 454–457, Piscataway, NJ, USA. IEEE Press

Chapter 7
Mining Source Code for Snippet Reuse

7.1 Overview

As already mentioned in the previous chapters, modern software development practices rely more and more on reuse. Systems are built using components found in software libraries and integrating them by means of small source code fragments, called *snippets*. As such, a large part of the development effort is spent on finding the proper snippets to perform the different envisioned tasks (e.g., read a CSV file, send a file over ftp, etc.) and integrate them into one's own source code. Using typical tools, such as search engines and programming forums for this task is far from optimal, as it requires leaving one's IDE to navigate through several online pages, in an attempt to comprehend the different ways to solve the problem at hand before selecting and integrating an implementation.

As a result, several methodologies have been proposed to address this challenge, most of which focus on the problems of *API usage mining* and *snippet mining*. API usage mining systems extract and present examples for specific library APIs [1–6]. Though effective, these systems are only focused on finding out how to use an API, without providing solutions in generic cases or in cases when determining which library to use is part of the question. Furthermore, several of them [1–3] return API call sequences instead of ready-to-use snippets.

On the other hand, generic snippet mining systems [7–9] employ code indexing mechanisms and thus include snippets for many different queries. Nevertheless, they also have important limitations. Concerning systems with local indexes [7], the quality and the diversity of their results are usually confined by the size of the index. Moreover, the retrieved snippets for all systems [7–9] are presented in the form of lists that do not allow easily distinguishing among different implementations (e.g., using different libraries to perform file management). The quality and the reusability of the results are also usually not evaluated. Finally, a common limitation in certain systems is that they involve some specialized query language, which may require additional effort by the developer.

© Springer Nature Switzerland AG 2020
T. Diamantopoulos and A. L. Symeonidis, *Mining Software Engineering Data for Software Reuse*, Advanced Information and Knowledge Processing,
https://doi.org/10.1007/978-3-030-30106-4_7

In this chapter, we design and develop *CodeCatch*, a system that receives queries in natural language, and employs the Google search engine to extract useful snippets from multiple online sources. Our system further evaluates the readability of the retrieved snippets, as well as their preference/acceptance by the developer community using information from online repositories. Moreover, CodeCatch performs clustering to group snippets according to their API calls, allowing the developer to first select the desired API implementation and subsequently choose which snippet to use.

7.2 State of the Art on Snippet and API Mining

As already mentioned, in this chapter we focus on systems that receive queries for solving specific programming tasks and recommend source code snippets suitable for reuse in the developer's source code. Some of the first systems of this kind, such as Prospector [10] or PARSEWeb [11], focused on the problem of finding a path between an input and an output object in source code. For Prospector [10], such paths are called jungloids and the resulting program flow is called a jungloid graph. The tool is quite effective for certain reuse scenarios and can also generate code. However, it requires maintaining a local database, which may easily become deprecated. A rather more broad solution was offered by PARSEWeb [11], which employed the Google Code Search Engine[1] and thus the resulting snippets were always up-to-date. Both systems, however, consider that the developer knows exactly which API objects to use, and he/she is only concerned with integrating them.

Another category of systems is those that generate API usage examples in the form of call sequences by mining client code (i.e., code using the API under analysis). One such system is MAPO [1], which employs *frequent sequence mining* in order to identify common usage patterns. As noted, however, by Wang et al. [2], MAPO does not account for the diversity of usage patterns, and thus outputs a large number of API examples, many of which are redundant. To improve on this aspect, the authors propose UP-Miner [2], a system that aims to achieve high coverage and succinctness. UP-Miner models client source code using graphs and mines frequent closed API call paths/sequences using the BIDE algorithm [12]. A screenshot of UP-Miner is shown in Fig. 7.1. The query in this case concerned the API call SQLConnection.Open, and, as one may see in this figure, the results include different scenarios, e.g., for reading (pattern 252) or for executing transactions (pattern 54), while the pattern is also presented graphically at the right part of the screen.

PAM [3] is another similar system that employs probabilistic machine learning to extract API call sequences, which are proven to be more representative than those of MAPO and UP-Miner. An interesting novelty of PAM [3] is the use of an automated evaluation framework based on handwritten usage examples by the developers of

---

[1] As already mentioned in the previous chapters, the Google Code Search Engine resided in http://www.google.com/codesearch, however, the service was discontinued in 2013.

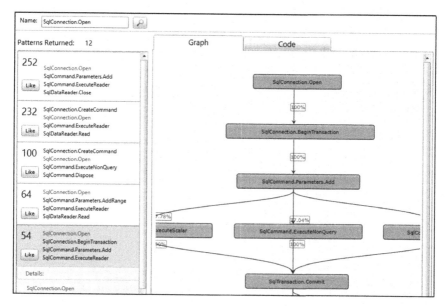

Fig. 7.1 Screenshot of UP-Miner [2]

the API under analysis. CLAMS [13] also employs the same evaluation process to illustrate how effective clustering can lead to patterns that are precise and at the same time diverse. The system also employs summarization techniques to produce snippets that are more readable and, thus, more easily reusable by the developer.

Apart from the aforementioned systems, which extract API call sequences, there are also approaches that recommend ready-to-use snippets. Indicatively, we refer to APIMiner [4], a system that performs code slicing to isolate useful API-relevant statements of snippets. Buse and Weimer [5] further employ path-sensitive data flow analysis and pattern abstraction techniques to provide more abstract snippets. Another important advantage of their implementation is that it employs clustering to group the resulting snippets into categories. A similar system in this aspect is eXoaD-ocs [6], as it also clusters snippets, however, using a set of semantic features proposed by the DECKARD code clone detection algorithm [14]. The system provides example snippets using the methods of an API and allows also receiving feedback from the developer. An example screenshot of eXoaDocs is shown in Fig. 7.2.

Though interesting, all of the aforementioned approaches provide usage examples for specific API methods, and do not address the challenge of choosing which library to use. Furthermore, several of these approaches output API call sequences, instead of ready-to-use solutions in the form of snippets. Finally, none of the aforementioned systems accept queries in natural language, which are certainly preferable when trying to formulate a programming task without knowing which APIs to use beforehand.

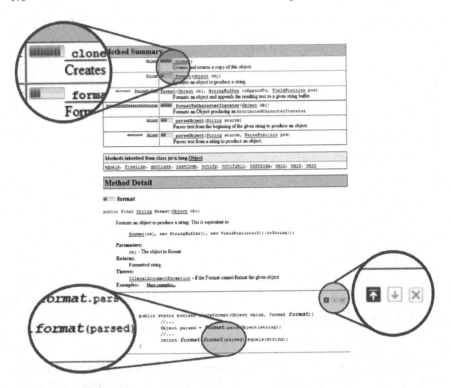

Fig. 7.2 Example screenshot of eXoaDocs [6]

To address the above challenges, several recent systems focus on generic snippets
and employ some type of processing for natural language queries. Such an example system is SnipMatch [7], the snippet recommender of the Eclipse IDE, which
also incorporates several interesting features, including variable renaming for easier
snippet integration. However, its index has to be built by the developer who has to provide the snippets and the corresponding textual descriptions. A relevant system also
offered as an Eclipse plugin is Blueprint [8]. Blueprint employs the Google search
engine to discover and rank snippets, thus ensuring that useful results are retrieved
for almost any query. An even more advanced system is Bing Code Search [9], which
employs the Bing search engine for finding relevant snippets, and further introduces
a multi-parameter ranking system for snippets as well as a set of transformations to
adapt the snippet to the source code of the developer.

The aforementioned systems, however, do not provide a choice of implementations to the developer. Furthermore, most of them do not assess the retrieved snippets
from a reusability perspective. This is crucial, as libraries that are preferred by developers typically exhibit high quality and good documentation, while they are obviously
supported by a larger community [15, 16]. In this chapter, we present CodeCatch,
a snippet mining system designed to overcome the above limitations. CodeCatch
employs the Google search engine in order to receive queries in natural language

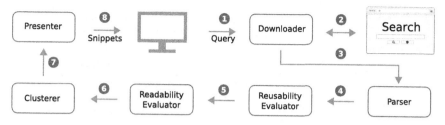

Fig. 7.3 CodeCatch system overview

and at the same time extract snippets from multiple online sources. As opposed to current systems, our tool assesses not only the quality (readability) of the snippets but also their reusability/preference by the developers. Furthermore, CodeCatch employs clustering techniques in order to group the snippets according to their API calls, and thus allows the developer to easily distinguish among different implementations.

7.3 CodeCatch Snippet Recommender

The architecture of CodeCatch is shown in Fig. 7.3. The input of the system is a query given in natural language to the Downloader module, which is then posted to the Google search engine, to extract code snippets from the result pages. Consequently, the Parser extracts the API calls of the snippets, while the Reusability Evaluator scores the snippets according to whether they are widely used/preferred by developers. Additionally, the readability of the snippets is assessed by the Readability Evaluator. Finally, the Clusterer groups the snippets according to their API calls, while the Presenter ranks them and presents them to the developer. These modules are analyzed in the following subsections.

7.3.1 Downloader

The downloader receives as input the query of the developer in natural language and posts it in order to retrieve snippets from multiple sources. An example query that will be used throughout this section is the query "How to read a CSV file". The downloader receives the query and augments it before issuing it in the Google search engine. Note that our methodology is programming language-agnostic; however, without loss of generality we focus in this work on the Java programming language. In order to ensure that the results returned by the search engine will be targeted to the Java language, the query augmentation is performed using the Java-related keywords java, class, interface, public, protected, abstract, final, static, import,

Fig. 7.4 Example snippet
for "How to read a CSV file"

```
String line = "";
BufferedReader br = null;
try {
    br = new BufferedReader(new FileReader("test.csv"));
    while((line = br.readLine()) != null) {
        String[] data = line.split(",");
    }
    br.close();
} catch (Exception e) {
    System.err.println("CSV file cannot be read: " + e);
}
```

if, for, void, int, long, and double. Similar lists of keywords can be constructed for
supporting other languages.

The URLs that are returned by Google are scraped using Scrapy.[2] In specific, upon
retrieving the top 40 web pages, we extract text from HTML tags such as *<pre>*
and *<code>*. Scraping from those tags allows us to gather the majority (around 92%
as measured) of code content from web pages. Apart from the code, we also collect
information relevant to the webpage, including the URL of the result, its rank at the
Google search results list, and the relative position of each code fragment inside the
page.

7.3.2 Parser

The snippets are then parsed using the sequence extractor described in [17], which
extracts the AST of each snippet and takes two passes over it, one to extract all
(non-generic) type declarations (including fields and variables), and one to extract
the method invocations (API calls). Consider, for example, the snippet of Fig. 7.4.
During the first pass, the parser extracts the declarations line: String, br: Buffere-
dReader, data: String[], and e: Exception. Then, upon removing the generic dec-
larations (i.e., literals, strings, and exceptions), the parser traverses the AST for a
second time and extracts the relevant method invocations, which are highlighted in
Fig. 7.4. The caller of each method invocation is replaced by its type (apart from
constructors for which types are already known), to finally produce the API calls
FileReader.__init__, BufferedReader.__init__, BufferedReader.readLine, and
BufferedReader.close, where __init__ signifies a call to a constructor.

Note that the parser is quite robust even in cases when the snippets are not com-
pilable, while it also effectively isolates API calls that are not related to a type (since
generic calls, such as close, would only add noise to the invocations). Finally, any

---

[2]https://scrapy.org/.

snippets not referring to Java source code and/or not producing API calls are dropped at this stage.

7.3.3 Reusability Evaluator

Upon gathering the snippets and extracting their API calls, the next step is to determine whether they are expected to be of use to the developer. In this context of *reusability*, we want to direct the developer toward what we call *common practice*, and, to do so, we make the assumption that snippets with API calls *commonly used* by other developers are more probable to be of (re)use. This is a reasonable assumption since answers to common programming questions are prone to appear often in the code of different projects. As a result, we designed the Reusability Evaluator by downloading a set of high-quality projects and determining the amount of reuse for the API calls of each snippet.

For this task, we have used the 1000 most popular Java projects of GitHub, as determined by the number of stars assigned. These were drawn from the index of AGORA (see Chap. 5). The rationale behind this choice of projects is highly intuitive and is also strongly supported by current research; popular projects have been found to exhibit high quality [15], while they contain reusable source code [18] and sufficient documentation [16]. As a result, we expect that these projects use the most effective APIs in a good way.

Upon downloading the projects, we construct a local index where we store their API calls, which are extracted using the Parser.[3] After that, we score each API call by dividing the number of projects in which it is present by the total number of projects. For the score of each snippet, we average over the scores of its API calls. Finally, the index also contains all qualified names so that we may easily retrieve them given a caller object (e.g., BufferedReader: java.io.BufferedReader).

7.3.4 Readability Evaluator

To construct a model for the readability of snippets, we used the publicly available dataset from [19] that contains 12,000 human judgements by 120 annotators on 100 snippets of code. We build our model as a binary classifier that assesses whether a code snippet is *more readable* or *less readable*. At first, for each snippet, we extract a set of features that are related to readability, including the average line length, the average identifier length, the average number of comments, etc. (see [19] for the full

---

[3]As a side note, this local index is only used to assess the reusability of the components; our search, however, is not limited within it (as is the case with other systems) as we employ a search engine and crawl multiple pages. To further ensure that our reusability evaluator is always up-to-date, we could rebuild its index along with the rebuild cycles of the index of AGORA.

Fig. 7.5 Example snippet
for "How to read a CSV file"
using Scanner

```
Scanner scanner = null;
try{
    scanner = new Scanner(new File("test.csv"));
    scanner.useDelimiter(",");
    while(scanner.hasNext()) {
        System.out.print(scanner.next() + " ");
    }
    scanner.close();
} catch (Exception e) {
    System.err.println("CSV file cannot be read: " + e);
}
```

list of features). After that, we train an AdaBoost classifier on the aforementioned
dataset. The classifier was built with decision trees as the base estimator, while the
number of estimators and the learning rate were set to 160 and 0.6, respectively.
We built our model using 10-fold cross-validation and the average F-measure for all
folds was 85%, indicating that it is effective enough for making a binary decision
concerning the readability of new snippets.

7.3.5 Clusterer

Upon scoring the snippets for both their reusability and their readability, the next
step is to cluster them according to the different implementations. In order to perform
clustering, we first have to extract the proper features from the retrieved snippets. A
simple approach would be to cluster the snippets by examining them as text documents; however, this approach would fail to distinguish among different implementations. Consider, for example, the snippet of Fig. 7.4 along with that of Fig. 7.5. If
we remove any punctuation and compare the two snippets, we may find out that more
than 60% of the tokens of the second snippet are also present in the first. The two
snippets, however, are quite different; they have different API calls and thus refer to
different implementations.

As a result, we decided to cluster code snippets based on their API calls. To do
so, we employ a Vector Space Model (VSM) to represent snippets as documents and
API calls as vectors (dimensions). Thus, at first, we construct a document for each
snippet based on its API calls. For example, the document for the snippet of Fig. 7.4
is "FileReader.__init__ BufferedReader.__init__ BufferedReader.readLine
Buf feredReader.close", while the document for that of Fig. 7.5 is "File.__init__
Scanner.__init__ Scanner.hasNext Scanner.next Scanner.close". After that,
we use a *tf-idf* vectorizer to extract the vector representation for each document. The
weight (vector value) of each term t in a document d is computed by the following
equation:

$$tfidf(t, d, D) = tf(t, d) \cdot idf(t, D) \tag{7.1}$$

where $tf(t, d)$ is the term frequency of term t in document d and refers to the appearances of the API call in the snippet, while $idf(t, D)$ is the inverse document frequency of term t in the set of all documents D, referring to how common the API call is in all the retrieved snippets. In specific, $idf(t, D)$ is computed as follows:

$$idf(t, D) = 1 + log \frac{1 + |D|}{1 + d_t} \tag{7.2}$$

where $|d_t|$ is the number of documents containing the term t, i.e., the number of snippets containing the relevant API call. The idf ensures that very common API calls (e.g., Exception.printStackTrace) are given low weights, so that they do not outweigh more decisive invocations.

Before clustering, we also need to define a distance metric that shall be used to measure the similarity between two vectors. Our measure of choice is the cosine similarity, which is defined for two document vectors d_1 and d_2 using the following equation:

$$cosine\_similarity(d_1, d_2) = \frac{d_1 \cdot d_2}{|d_1| \cdot |d_2|} = \frac{\sum_1^N w_{t_i, d_1} \cdot w_{t_i, d_2}}{\sum_1^N w_{t_i, d_1}^2 \cdot \sum_1^N w_{t_i, d_2}^2} \tag{7.3}$$

where w_{t_i, d_1} and w_{t_i, d_2} are the tf-idf scores of term t_i in documents d_1 and d_1, respectively, and N is the total number of terms.

We select K-Means as our clustering algorithm, as it is known to be effective in text clustering problems similar to ours [20]. The algorithm, however, still has an important limitation as it requires as input the number of clusters. Therefore, to automatically determine the best value for the number of clusters, we employ the silhouette metric. The silhouette was selected as it encompasses both the similarity of the snippets within the cluster (cohesion) and their difference with the snippets of other clusters (separation). As a result, we expect that using the value of the silhouette as an optimization parameter should yield clusters that clearly correspond to different implementations. We execute K-Means for 2–8 clusters, and each time we compute the value of the silhouette coefficient for each document (snippet) using the following equation:

$$silhouette(d) = \frac{b(d) - a(d)}{max(a(d), b(d))} \tag{7.4}$$

where $a(d)$ is the average distance of document d from all other documents in the same cluster, while $b(d)$ is computed by measuring the average distance of d from the documents of each of the other clusters and keeping the lowest one of these values (each corresponding to a cluster). For both parameters, the distances between documents are measured using Eq. (7.3). Finally, the silhouette coefficient for a cluster is given as the mean of the silhouette values of its snippets, while the

Fig. 7.6 Silhouette score of
different number of clusters
for the example query "How
to read a CSV file"

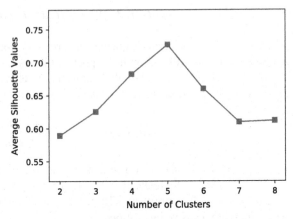

Fig. 7.7 Silhouette score of
each cluster when grouping
into 5 clusters for the
example query "How to read
a CSV file"

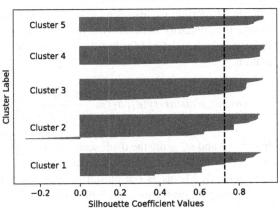

total silhouette for all clusters is given by averaging over the silhouette values of all
snippets.

We present an example silhouette analysis for the query "How to read a CSV
file". Figure 7.6 depicts the silhouette score for 2–8 clusters, where it is clear that the
optimal number of clusters is 5.

Furthermore, the individual silhouette values for the documents (snippets) of
the five clusters are shown in Fig. 7.7, and they also confirm that the clustering is
effective as most samples exhibit high silhouette and only a few have marginally
negative values.

7.3.6 Presenter

The presenter is the final module of CodeCatch and handles the ranking and the
presentation of the results. We have developed the CodeCatch prototype as a web

Fig. 7.8 Screenshot of CodeCatch for query "How to read a CSV file", depicting the top three clusters

application. When the developer inserts a query, he/she is first presented with the clusters that correspond to different implementations for the query.

An indicative view of the first three clusters containing CSV file reading implementations is shown in Fig. 7.8. The proposed implementations of the clusters involve the BufferedReader API (e.g., as in Fig. 7.4), the Scanner API (e.g., as in Fig. 7.5), and the Java CSV reader API.[4] The clusters are ordered according to their API reusability score, which is the average of the score of each of their snippets, as defined in Sect. 7.3.3. For each cluster, CodeCatch provides the five most frequent API imports and the five most frequent API calls, to aid the developer to distinguish among the different implementations. In cases where imports are not present in the snippets, they are extracted using the index created in Sect. 7.3.3.

Upon selecting to explore a cluster, the developer is presented with a list of its included snippets. The snippets within a cluster are ranked according to their API reusability score, and in cases of equal scores according to their distance from the cluster centroid (computed using Eq. (7.3)). This ensures that the most common usages of a specific API implementation are higher on the list. Furthermore, for each snippet, CodeCatch provides useful information, as shown in Fig. 7.9, including its reusability score (*API Score*), its distance from the centroid, its readability (either Low or High), the position of its URL in the results of the Google search engine and its order inside the URL, its number of API calls, and its number of lines of code. Finally, apart from immediately reusing the snippet, the developer has the option to isolate only the code that involves its API calls, while he/she can also check the webpage from which the snippet was retrieved.

7.4 Example Usage Scenario

In this section, we provide an example usage scenario of CodeCatch based on the scenario described in the previous chapter. In specific, we build on top of the duplicate file check application that was developed using our component reuse methodology.

---

[4]https://gist.github.com/jaysridhar/d61ea9cbede617606256933378d71751.

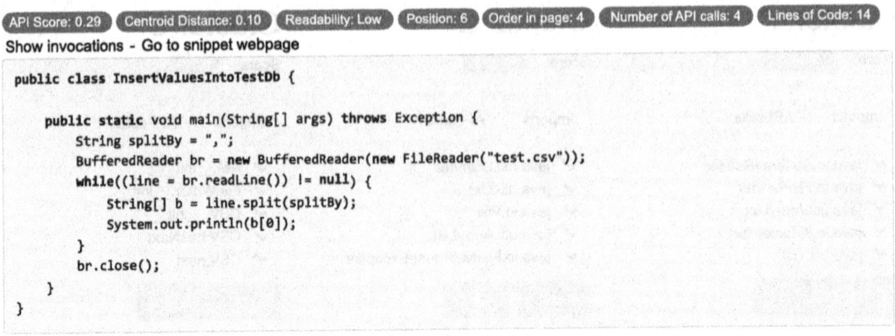

API Score: 0.29 Centroid Distance: 0.10 Readability: Low Position: 6 Order in page: 4 Number of API calls: 4 Lines of Code: 14

Show invocations - Go to snippet webpage

```
public class InsertValuesIntoTestDb {

    public static void main(String[] args) throws Exception {
        String splitBy = ",";
        BufferedReader br = new BufferedReader(new FileReader("test.csv"));
        while((line = br.readLine()) != null) {
            String[] b = line.split(splitBy);
            System.out.println(b[0]);
        }
        br.close();
    }
}
```

Fig. 7.9 Screenshot of CodeCatch for query "How to read a CSV file", depicting an example snippet

Table 7.1 API clusters of CodeCatch for query "How to upload file to FTP"

ID	Library	API	Score (%)
1	Apache commons	org.apache.commons.net.ftp.*	83.57
2	Jscape	com.jscape.inet.ftp.*	11.39
3	EnterpriseDT edtFTPj	com.enterprisedt.net.ftp.*	4.78
4	Spring framework	org.springframework.integration.ftp.*	0.26

In this case, we assume that the developer has built the relevant application, which scans the files of a given folder and checks their MD5 checksums before adding them to a given path. Let us now assume that the next step involves uploading these files to an FTP server.

FTP file handling is typically a challenge that can be confronted using external libraries/frameworks, therefore the process of writing code to perform this task is well suited for CodeCatch. Thus, the developer, who does not necessarily know about any components and/or APIs, initially issues the simple textual query "How to upload file to FTP". In this case, CodeCatch returns four result clusters, each linked to a different API (and, thus, a different framework). The returned API clusters are summarized in Table 7.1 (the API column is returned by the service, while the library column is provided here for clarity).

As one may easily observe, there are different equivalent implementations for uploading files to an FTP server. The developer could, of course, choose any of them, depending on whether he/she already is familiar with any of these libraries. However, in most scenarios (and let us assume also in our scenario), the developer would select to use the most highly preferred implementation, which in this case is the implementation offered by Apache Commons. Finally, upon selecting to enter the first cluster, the developer could review the snippets and select one that fits his/her purpose, such as the one shown in Fig. 7.10.

```
String ftpUrl = "ftp://%s:%s@%s/%s;type=i";
String host = "www.myserver.com";
String user = "tom";
String pass = "secret";
String filePath = "E:/Work/Project.zip";
String uploadPath = "/MyProjects/archive/Project.zip";

ftpUrl = String.format(ftpUrl, user, pass, host, uploadPath);
System.out.println("Upload URL: " + ftpUrl);

try {
    URL url = new URL(ftpUrl);
    URLConnection conn = url.openConnection();
    OutputStream outputStream = conn.getOutputStream();
    FileInputStream inputStream = new FileInputStream(filePath);
    byte[] buffer = new byte[BUFFER_SIZE];
    int bytesRead = −1;
    while ((bytesRead = inputStream.read(buffer)) != −1) {
        outputStream.write(buffer, 0, bytesRead);
    }
    inputStream.close();
    outputStream.close();
    System.out.println("File uploaded");
} catch (IOException ex) {
    ex.printStackTrace();
}
```

Fig. 7.10 Example snippet for "How to upload file to FTP" using Apache Commons

7.5 Evaluation

7.5.1 Evaluation Framework

Comparing CodeCatch with similar approaches has not been performed in a straightforward manner, as several of them focus on mining single APIs, while others are not maintained and/or are not publicly available. Our focus is mainly on the merit of reuse for results, and the system that is most similar to ours is Bing Code Search [9], however, it targets the C# programming language. Hence, we have decided to perform a reusability-related evaluation against the Google search engine on a dataset of common queries shown in Table 7.2.

The purpose of our evaluation is twofold; we wish not only to assess whether the snippets of our system are relevant, but also to determine whether the developer

Table 7.2 Statistics of the queries used as evaluation dataset

ID	Query	Clusters	Snippets
1	How to read CSV file	5	76
2	How to generate MD5 hash code	5	65
3	How to send packet via UDP	5	34
4	How to split string	4	22
6	How to upload file to FTP	4	31
5	How to send email	5	79
7	How to initialize thread	6	51
8	How to connect to a JDBC database	5	42
9	How to read ZIP archive	6	82
10	How to play audio file	6	45

can indeed more easily find snippets for all different APIs relevant to a query. Thus, at first, we annotate the retrieved snippets for all the queries as relevant and non-relevant. To maintain an objective and systematic outlook, the annotation procedure was performed without any knowledge on the ranking of the snippets. For the same reason, the annotation was kept as simple as possible; snippets were marked as relevant if and only if their code covers the functionality described by the query. That is, for the query, "How to read CSV file", any snippets used to read a CSV file were considered relevant, regardless of their size or complexity, and of any libraries involved, etc.

As already mentioned, the snippets are assigned to clusters, where each cluster involves different API usages and thus corresponds to a different implementation. As a result, we have to assess the relevance of the results per cluster, hence assuming that the developer would first select the desired implementation and then navigate into the cluster. To do so, we compare the snippets of each cluster (i.e., of each implementation) to the results of the Google search engine. CodeCatch clusters already provide lists of snippets, while for Google, we construct one by assuming that the developer opens the first URL, subsequently examines the snippets of this URL from top to bottom, then he/she opens the second URL, etc.

When assessing the results of each cluster, we wish to find snippets that are relevant not only to the query but also to the corresponding API usages. As a result, for the assessment of each cluster, we further annotate the results of both systems to consider them relevant when they are also part of the corresponding implementation. This, arguably, produces less effective snippet lists for the Google search engine, however, note that our purpose is not to challenge the results of the Google search engine in terms of relevance to the query, but rather to illustrate how easy or hard it is for the developer to examine the results and isolate the different ways of answering his/her query.

For each query, upon having constructed the lists of snippets for each cluster and for Google, we compare them using the *reciprocal rank* metric. This metric was selected as it is commonly used to assess information retrieval systems in general and also systems similar to ours [9]. Given a list of results, the reciprocal rank for a query is computed as the inverse of the rank of the first relevant result. For example, if the first relevant result is in the first position, then the reciprocal rank is $1/1 = 1$, if the result is in the second position, then the reciprocal rank is $1/2 = 0.5$, etc.

7.5.2 Evaluation Results

Figure 7.11 depicts the reciprocal rank of CodeCatch and Google for the snippets corresponding to the three most popular implementations for each query. At first, interpreting this graph in terms of the relevance of the results indicates that both systems are very effective. In specific, if we consider that the developer would require a relevant snippet regardless of the implementation, then for most queries, both CodeCatch and Google produce a relevant result in the first position (i.e., reciprocal rank equal to 1).

If, however, we focus on all different implementations for each query, we can make certain interesting observations. Consider, for example, the first query ("How to read a CSV file"). In this case, if the developer requires the most popular BufferedReader implementation (I1), both CodeCatch and Google output a relevant snippet in the first position. Similarly, if one wished to use the Scanner implementation (I2) or the Java CSV reader (I3), our system would return a ready-to-use snippet in the top of the second cluster or in the second position of the third cluster (i.e., reciprocal rank equal to 0.5). On the other hand, using Google would require examining more results (3 and 50 results for I2 and I3, respectively, as the corresponding reciprocal ranks are equal to 0.33 and 0.02, respectively). Similar conclusions can be drawn for most queries.

Another important point of comparison of the two systems is whether they return the most popular implementations at the top of their list. CodeCatch is clearly more effective than Google in this aspect. Consider, for example, the sixth query; in this case, the most popular implementation is found in the third position of Google, while the snippet found in its first position corresponds to a less popular implementation. This is also clear in several other queries (i.e., queries 2, 3, 5, 7). Thus, one could argue that CodeCatch does not only provide the developer with all different API implementations for his/her query but also further aids him/her to select the most popular of these implementations, which is usually the most preferable.

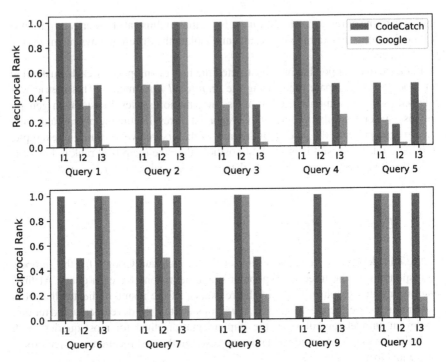

Fig. 7.11 Reciprocal Rank of CodeCatch and Google for the three most popular implementations (I1, I2, I3) of each query

7.6 Conclusion

In this chapter, we proposed a system that extracts snippets from online sources and further assesses their readability as well as their reusability based on the preference of developers. Moreover, our system, CodeCatch, provides a comprehensive view of the retrieved snippets by grouping them into clusters that correspond to different implementations. This way, the developer can select among possible solutions to common programming queries even in cases when he/she does not know which API to use beforehand.

Future work on CodeCatch lies in several directions. At first, we may extend our ranking scheme to include the position of each snippet's URL in the Google results, etc. Furthermore, snippet summarization can be performed using information from the clustering (e.g., removing statements that appear in few snippets within a cluster). Finally, an interesting idea would be to conduct a developer study in order to further assess CodeCatch for its effectiveness in retrieving useful source code snippets.

References

1. Xie T, Pei J (2006) MAPO: mining API usages from open source repositories. In: Proceedings of the 2006 international workshop on mining software repositories, MSR '06, pp 54–57, New York, NY, USA. ACM
2. Wang J, Dang Y, Zhang H, Chen K, Xie T, Zhang D (2013) Mining succinct and high-coverage API usage patterns from source code. In: Proceedings of the 10th working conference on mining software repositories, pp 319–328, Piscataway, NJ, USA. IEEE Press
3. Fowkes J, Sutton C (2016) Parameter-free probabilistic API mining across GitHub. In: Proceedings of the 2016 24th ACM SIGSOFT international symposium on foundations of software engineering, pp 254–265, New York, NY, USA. ACM
4. Montandon JE, Borges H, Felix D, Valente MT (2013) Documenting APIs with examples: lessons learned with the APIMiner platform. In: 2013 20th working conference on reverse engineering (WCRE), pp 401–408, Piscataway, NJ, USA. IEEE Computer Society
5. Buse RPL, Weimer W (2012) Synthesizing API usage examples. In: Proceedings of the 34th international conference on software engineering, ICSE '12, pp 782–792, Piscataway, NJ, USA. IEEE Press
6. Kim J, Lee S, Hwang SW, Kim S (2010) Towards an intelligent code search engine. In: Proceedings of the Twenty-Fourth AAAI conference on artificial intelligence, AAAI'10, pp 1358–1363, Palo Alto, CA, USA. AAAI Press
7. Wightman D, Ye Z, Brandt J, Vertegaal R (2012) SnipMatch: using source code context to enhance snippet retrieval and parameterization. In: Proceedings of the 25th annual ACM symposium on user interface software and technology, UIST '12, pp 219–228, New York, NY, USA. ACM
8. Brandt J, Dontcheva M, Weskamp M, Klemmer SR (2010) Example-centric programming: integrating web search into the development environment. In: Proceedings of the SIGCHI conference on human factors in computing systems, CHI '10, pp 513–522, New York, NY, USA. ACM
9. Wei Y, Chandrasekaran N, Gulwani S, Hamadi Y (2015) Building bing developer assistant. Technical Report MSR-TR-2015-36, Microsoft Research
10. Mandelin D, Lin X, Bodík R, Kimelman D (2005) Jungloid mining: helping to navigate the API jungle. SIGPLAN Not 40(6):48–61
11. Thummalapenta S, Xie T (2007) PARSEWeb: a programmer assistant for reusing open source code on the web. In: Proceedings of the 22nd IEEE/ACM international conference on automated software engineering, ASE '07, pp 204–213, New York, NY, USA. ACM
12. Wang J, Han J (2004) BIDE: efficient mining of frequent closed sequences. In: Proceedings of the 20th international conference on data engineering, ICDE '04, pp 79–90, Washington, DC, USA. IEEE Computer Society
13. Katirtzis N, Diamantopoulos T, Sutton C (2018) Summarizing software API usage examples using clustering techniques. In: 21th international conference on fundamental approaches to software engineering, pp 189–206, Cham. Springer International Publishing
14. Jiang L, Misherghi G, Su Z, Glondu S (2007) DECKARD: scalable and accurate tree-based detection of code clones. In: Proceedings of the 29th international conference on software engineering, ICSE '07, pp 96–105, Washington, DC, USA. IEEE Computer Society
15. Papamichail M, Diamantopoulos T, Symeonidis AL (2016) User-perceived source code quality estimation based on static analysis metrics. In: Proceedings of the 2016 IEEE international conference on software quality, reliability and security, QRS, pp 100–107, Vienna, Austria
16. Aggarwal K, Hindle A, Stroulia E (2014) Co-evolution of project documentation and popularity within Github. In: Proceedings of the 11th working conference on mining software repositories, MSR 2014, pp 360–363, New York, NY, USA. ACM
17. Diamantopoulos T, Symeonidis AL (2015) Employing source code information to improve question-answering in stack overflow. In: Proceedings of the 12th working conference on mining software repositories, MSR '15, pp 454–457, Piscataway, NJ, USA. IEEE Press

18. Dimaridou V, Kyprianidis A-C, Papamichail M, Diamantopoulos T, Symeonidis A (2017) Towards modeling the user-perceived quality of source code using static analysis metrics. In: Proceedings of the 12th international conference on software technologies - Volume 1, ICSOFT, pp 73–84, Setubal, Portugal. INSTICC, SciTePress

19. Buse RPL, Weimer WR (2010) Learning a metric for code readability. IEEE Trans Softw Eng 36(4):546–558

20. Aggarwal CC, Zhai C (2012) A survey of text clustering algorithms, pp 77–128. Springer, Boston

Chapter 8
Mining Solutions for Extended Snippet Reuse

8.1 Overview

In the previous chapter, we analyzed the challenges a developer faces when trying to find solutions to programming queries online. We focused on API-agnostic usage queries that are issued by the developer in order to find implementations that satisfy the required functionality. In this chapter, we focus on a challenge that follows from snippet mining, that is, of finding out whether a snippet chosen (or generally written) by the developer is correct and also optimal for the task at hand. In specific, we assume that the developer has either written a source code snippet or found and integrated it in his/her source code, and yet the snippet does not execute as expected. In such a scenario, a common process followed involves searching online, possibly in question-answering communities like Stack Overflow, in order to probe on how other users have handled the problem at hand, or whether they have already found a solution. In case that an existing answer to the problem is not found, the developer usually submits a new question post, hoping that the community will respond to his/her aid.

Submitting a new question post in Stack Overflow requires several steps; the developer has to form a question with a clear title, further explain the problem in the question body, isolate any relevant source code fragments, and possibly assign tags to the question (e.g., "swing" or "android") to draw the attention of community members that are familiar with the specific technologies/frameworks. In most cases, however, the first step before forming a question post is to find out whether the same question (or a similar one) is already posted. To do so, one can initially perform a search on question titles. Further improving the results requires including tags or question bodies. Probably the most effective query includes the title, the tags, and the body, i.e., the developer has to form a complete question post. This process is obviously complex (for the newcomer) and time-consuming; for example, the source code of the problem may not be easily isolated, or the developer may not know which tags to use.

© Springer Nature Switzerland AG 2020
T. Diamantopoulos and A. L. Symeonidis, *Mining Software Engineering Data for Software Reuse*, Advanced Information and Knowledge Processing,
https://doi.org/10.1007/978-3-030-30106-4_8

In this chapter, we explore the problem of finding similar question posts in Stack Overflow, given different types of information. Apart from titles, tags, or question bodies, we explore whether a developer could be able to search for similar questions using source code fragments, thus not requiring to fully form a question. Our main contribution is a question similarity scheme which can be used to recommend similar question posts to users of the community trying to find a solution to their problem [1]. Using our scheme, developers shall be able to check whether their snippets (either handwritten or reused from online sources) are good choices for the functionality that has to be covered. As a side note, since this scheme could be useful for identifying similar questions, it may also prove useful to community members or moderators so that they identify linked or duplicate questions, or even so that they try to find similar questions to answer and thus contribute to the community.

8.2 Data Collection and Preprocessing

Our methodology is applied to the official data dump of Stack Overflow as of September 26, 2014, provided by [2]. We have used Elasticsearch [3] to store the data for our analysis, since it provides indexing and supports the execution of fast queries on large chunks of data. Storage in Elasticsearch is simple; any record of data is a *document*, documents are stored inside *collections*, and collections are stored in an *index* (see Chap. 5 for more information on Elasticsearch-based architectures). We have defined a *posts* collection, which contains the Java questions of Stack Overflow (determined by the "java" tag).[1]

8.2.1 Extracting Data from Questions

An example Stack Overflow question post is depicted in Fig. 8.1. The main elements of the post are: the title (shown at the top of Fig. 8.1), the body, and the tags (at the bottom of Fig. 8.1). These three elements are stored and indexed for each question post in our Elasticsearch server to allow for retrieving the most relevant questions. The title and the body are stored as text, while the tags are stored as a list of keywords. Since the body is in html, we have also employed an html parser to extract the source code snippets for each question.

---

[1] Although our methodology is mostly language-agnostic, we have focused on Java Stack Overflow questions to demonstrate our proof of concept.

8.2.2 *Storing and Indexing Data*

This subsection discusses how the elements of each post are indexed in Elasticsearch. As already noted in Chap. 5, Elasticsearch not only stores the data but also creates and stores *analyzed* data for each field, so that the index can be searched in an efficient manner.

The title of each question post is stored in the index using the *standard analyzer* of Elasticsearch. This analyzer comprises a tokenizer that splits the text into tokens (i.e., lowercase words) and a token filter that removes certain common stop words (in our case the common English stop words). The tags are stored in an array field without further analysis, since they are already split and in lowercase. The body of each question post is analyzed using an *html analyzer*, which is similar to the standard analyzer, however, it first removes the html tags using the html_strip character filter of Elasticsearch. Titles, tags, and question bodies are stored under the fields "Title", "Tags", and "Body", respectively.

Concerning the snippets extracted from the question bodies, we had to find a representation that describes not only their terms (which are already included in the "Body" field), but also their structure. Source code representation is an interesting problem for which several solutions can be proposed. Although complex representations, such as the snippet's Abstract Syntax Tree (AST) or its program dependency graph, are rich in information, they cannot be used in the case of incomplete snippets such as the one in Fig. 8.1. For example, extracting structural information from the AST and using matching heuristics between classes and methods, as in [4], is not possible since we may not have information about classes and methods (e.g., inheritance, declared variables, etc.).

How to iterate through all buttons of grid layout?

I have a 2x2 grid of buttons

```
JFrame frame = new JFrame("myframe");
JPanel panel = new JPanel();
Container pane = frame.getContentPane();
GridLayout layout = new GridLayout(2,2);
panel.setLayout(layout);
panel.add(upperLeft);
panel.add(upperRight);
panel.add(lowerLeft);
panel.add(lowerRight);
pane.add(panel);
```

where upperLeft, upperRight, etc. are buttons. How can I iterate through all of the buttons?

java   swing   jbutton

Fig. 8.1 Example stack overflow question post

Table 8.1 Example lookup
table for the snippet of
Fig. 8.1

Variable	Type
frame	JFrame
panel	JPanel
pane	Container
layout	GridLayout

Interesting representations can be traced in the API discovery problem, where several researchers have used sequences to represent the snippets [5, 6]. However, as already noted in the previous chapter, these approaches concern how to call an API function, thus they are not applicable to the broader problem of snippet similarity. Other approaches include mining types from the snippet [7, 8] or tokenizing the code and treating it as a bag of words, perhaps also considering the entropy of any given term [9]. In any case, though, these representations result in loss of structural information. Thus, we had to find a representation that uses both the types and the structure of the snippets.

The representation we propose is a simple sequence, which is generated by three source code instruction types: assignments (AM), functions calls (FC), and class instantiations (CI). We deviate from complex representations, as in [10], since incomplete snippets may not contain all the declarations. For example, in the snippet of Fig. 8.1, we do not really know anything about the "upperLeft" object; it could be a field or a class. Our method operates on the AST of each snippet, extracted using the Eclipse compiler. In specific, we use the parser described in [1] that has already been used for our methodology in the previous chapter and is robust enough for different scenarios. Upon parsing the AST and extracting all individual commands, we take two passes over them (which are adequate, since the code is sequential). In the first pass, we extract all the declarations from the source code (including classes, fields, methods, and variables) and create a lookup table. The lookup table for the snippet of Fig. 8.1 is shown in Table 8.1. As shown in this table, the snippet includes four variables.

Upon extracting the types, the second pass creates a sequence of commands for the snippet. For example, the command "pane = frame.getContentPane()" provides an item FC_getContentPane. After that, we further refine this sequence item by substituting the name of the variable or function (in this case getContentPane) by its type (in this case Container) using also the lookup table when required. If no type can be determined, the type void is assigned. The sequence for the snippet of Fig. 8.1 is shown in Fig. 8.2.

This representation is also added to our Elasticsearch index in the "Snippets" field, as an ordered list.

Fig. 8.2 Example sequence
for the snippet of Fig. 8.1

Cl_JFrame
Cl_JPanel
FC_Container
Cl_GridLayout
FC_void
FC_void
FC_void
FC_void
FC_void
FC_void

8.3 Methodology

This section discusses our methodology, illustrating how text, tags, and snippets can be used to find similar questions. A query on the index may contain one or more of the "Title", "Tags", "Body", and "Snippets" fields. When searching for a question post, we calculate a score that is the average among the scores for each of the provided fields. The score for each field is normalized in the [0, 1] range, so that the significance of all fields is equal. The following subsections illustrate how the scores are computed for each field.

8.3.1 Text Matching

The similarity between two segments of text, either titles or bodies, is computed using the *term frequency-inverse document frequency (tf-idf)*. Upon splitting the text/document into tokens/terms (which is already handled by Elasticsearch), we use the vector space model, where each term is the value of a dimension of the model. The frequency of each term in each document (tf) is computed as the square root of the number of times the term appears in the document, while the inverse document frequency (idf) is the logarithm of the total number of documents divided by the number of documents that contain the term. The final score between two documents is computed using the cosine similarity between them:

$$score(d_1, d_2) = \frac{d_1 \cdot d_2}{|d_1| \cdot |d_2|} = \frac{\sum_1^N w_{t_i,d_1} \cdot w_{t_i,d_2}}{\sum_1^N w_{t_i,d_1}^2 \cdot \sum_1^N w_{t_i,d_2}^2} \tag{8.1}$$

where d_1, d_2 are the two documents, and w_{t_i,d_j} is the tf-idf score of term t_i in the document d_j.

8.3.2 Tag Matching

Since the values of "Tags" are unique, we can compute the similarity between "Tags" values by handling the lists as sets. Thus, we define the similarity as the *Jaccard index* between the sets. Given two sets, T_1 and T_2, their Jaccard index is the size of their intersection divided by the size of their union:

$$J(T_1, T_2) = \frac{|T_1 \cap T_2|}{|T_1 \cup T_2|} \qquad (8.2)$$

Finally, note that we exclude the tag "java" given the selected dataset, as it is not useful for the scoring.

8.3.3 Snippet Matching

Upon having extracted sequences from snippets (see Sect. 8.2.2), we define a similarity metric between two sequences based on their *Longest Common Subsequence (LCS)*. Given two sequences S_1 and S_2, their LCS is defined as the longest subsequence that is common to both sequences. Note that a subsequence is not required to contain consecutive elements of any sequence. For example, given two sequences $S_1 = [A, B, D, C, B, D, C]$ and $S_2 = [B, D, C, A, B, C]$, their LCS is $LCS(S_1, S_2) = [B, D, C, B, C]$. The LCS problem for two sequences S_1 and S_2 can be solved using dynamic programming. The computational complexity of the solution is $O(m \times n)$, where m and n are the lengths of the two sequences [11], so the algorithm is quite fast.

Upon having found the LCS between two snippet sequences S_1 and S_2, the final score for the similarity between them is computed by the following equation:

$$score(S_1, S_2) = 2 \cdot \frac{|LCS(S_1, S_2)|}{|S_1| + |S_2|} \qquad (8.3)$$

Since the length of the LCS is always smaller than the length of the smallest sequence, the fraction of it divided by the sum of sequence lengths is in the range $[0, 0.5]$. Hence, the score of the above equation lies in the range $[0, 1]$.

For example, if we calculate the score between the sequence of Fig. 8.2 and that of Fig. 8.3, the length of the LCS is 5, and the lengths of the two sequences are 10 and 6. Thus, the score between the two sequences, using Eq. (8.3), is 0.625.

8.4 Example Usage Scenario

To better highlight the potential of the methodology outlined in this chapter, we provide an example usage scenario, which is built on top of the ones discussed in

Cl_JFrame
FC_Container
AM_float
Cl_GridLayout
FC_void
FC_void

Fig. 8.3 Example sequence extracted from a source code snippet

Table 8.2 Example stack overflow questions that are similar to "uploading to FTP using java"

ID	Title	Score
1	How to resume the interrupted File upload in ftp using Java?	0.668
2	Testing file uploading and downloading speed using FTP	0.603
3	URLConnection slow to call getOutputStream using an FTP url	0.596
4	How to delete file from ftp server using Java?	0.573
5	Storing file name to database after uploading to ftp server	0.558
6	Copy all directories to server by FTP using Java	0.553
7	Getting corrupted JPG while uploading using FTP connection?	0.526
8	How to give FTP address in Java?	0.518
9	Can not connect to FTP server from heroku using apache client	0.501
10	FTP site works but I'm unable to connect from Java program	0.467

Chaps. 6 and 7. In specific, using our component reuse methodology (Chap. 6), the developer has built a duplicate file check application that scans the files of a folder and checks their MD5 checksums. Any unique files are then uploaded to an FTP server. The implementation for connecting to an FTP server and uploading files was supported using our snippet mining methodology (Chap. 7). The code snippet for the FTP implementation is presented in Fig. 7.10.

Assuming the developer has already integrated the snippet in his/her source code, his/her course of action is now quite diverse. For instance, it may be useful to find code for deleting some files erroneously placed on the FTP server or to find code for resuming the upload in case of some interruption or even to find code for debugging the application if there is some failure. What the developer could do in any of these cases is to build a question and post it on Stack Overflow, waiting for some useful answer.

As, in fact, the snippet of Fig. 7.10 is part of a Stack Overflow link, which contains the question "Uploading to FTP using Java" that has the tags "java", "ftp", and "uploading", we might safely assume that the developer would provide similar input. Using our approach, the formed query by the developer could be issued as input to the methodology defined in Sect. 8.3. In this case, the retrieved similar questions could provide valuable input for any of the aforementioned tasks. In specific, by issuing the query, the 10 most relevant question posts are shown in Table 8.2.

Fig. 8.4 Example code
snippet from stack overflow
post about "how to delete file
from ftp server using Java?"

```
FTPClient client = new FTPClient();
client.connect(host, port);
client.login(loginname, password);
client.deleteFile(fileNameOnServer);
client.disconnect();
```

The results are quite probable to be useful. So, for instance, if the developer wanted to issue the query in order to find out how to also delete files via FTP (instead of only adding them), he/she would be presented with an answer prompting him/her to employ the Apache FTPClient as in Fig. 8.4.

8.5 Evaluation

Evaluating the similarity of question posts in Stack Overflow is difficult, since the data are not annotated for this type of tests. We use the presence of a link between two questions (also given by [2]) as an indicator of whether the questions are similar. Although this assumption seems rational, the opposite assumption, i.e., that any non-linked questions are not similar, is not necessarily true. In fact, the Stack Overflow data dump has approximately 700,000 Java question posts, out of which roughly 300,000 have snippets that result in sequences (i.e., snippets with at least one assignment, function call, or class instantiation), and on average each of these posts has 0.98 links. In our scenario, however, the problem is formed as a search problem, so the main issue is to detect whether the linked (and thus similar) questions can indeed be retrieved.

In our evaluation, we have included only questions with at least 5 links (including duplicates which are also links according to the Stack Overflow data schema).[2] For performance reasons, all queries are at first performed on the question title, and then the first 1000 results are searched again using both title and any other fields, all in an Elasticsearch *bool* query. We used eight settings (given the title is always provided, the settings are all the combinations of tags, body, and snippets) to evaluate our methodology. For each question we keep the first 20 results, assuming this is the maximum a developer would examine (using other values had similar results).

The diagram of Fig. 8.5 depicts the average percentage of relevant results (compared to the total number of relevant links for each question) in the first 20 results of a query, regardless of the ranking.

These results involve all questions of the dataset that have source code snippets (approximately 300,000), even if the sequence extracted from them has length equal

---

[2]Our methodology supports all questions, regardless of the number of links. However, in the context of our evaluation, we may assume that questions with fewer links may be too localized and/or may not have similar question posts.

Fig. 8.5 Percentage of relevant results in the first 20 results of each query, evaluated for questions with snippet sequence length larger than or equal to 1

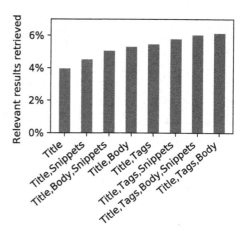

to 1 (having sequence length equal to 0 means no comparison can be made). Note that although the results for all settings in Fig. 8.5 are below 10%, this is actually a shortcoming of the dataset. Many more of the retrieved results may be relevant, however they are not linked.

As also shown in Fig. 8.5, when snippets are used in accordance with titles and/or tags, the retrieved questions are more relevant. However, when the body of a post is used, the use of snippets is probably redundant or even non-preferable. This is expected since the diagram of Fig. 8.5 involves many small snippets that result in sequences that are hard to compare. In these cases, supplying the mechanism with more (or larger) snippets, or even adding some text to the body is preferred.

To assert the validity of these claims and further discuss the effectiveness of our approach, we performed two more tests (results depicted in Figs. 8.6a and 8.6b). As in Fig. 8.5, the diagrams depict the percentage of relevant results in the first 20 results of a query, however, for snippets with sequence length larger than or equal to 3 and 5 (approximately 200,000 and 150,000 question posts respectively), for Figs. 8.6a and 8.6b, respectively. The effectiveness of using code snippets to find similar questions is quite clear from these two diagrams. As expected, using all available sources of information, i.e., question title, tags, body, and snippets, is optimal in both cases. Our snippet matching approach seems to be quite effective in retrieving relevant questions, even when the question body is not used.

Furthermore, when the provided snippets are larger and thus result in longer sequences, as in Fig. 8.6b, using them is more effective than using the whole question body. Another decisive piece of information that seems to affect the relevance of the results in all three evaluation scenarios is the use of tags. We notice that using tags and snippets is almost as effective as writing the whole question body. Thus, the developer could write a title and some tags for his/her question and then post a relevant piece of his/her code without requiring to fully formulate a question.

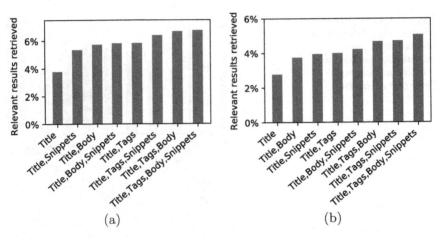

(a)   (b)

Fig. 8.6 Percentage of relevant results in the first 20 results of each query, evaluated for questions with snippet sequence length larger than or equal to **a** 3 and **b** 5

8.6 Conclusion

In this chapter, we have explored the problem of finding relevant questions in Stack Overflow. We have designed a similarity scheme which exploits not only the title, tags, and body of a question, but also its source code snippets. The results of our evaluation indicate that snippets are a valuable source of information for detecting links or for determining whether a question is already posted. As a result, using our system, the developer is able to easily form a question post even using only an already existing snippet in order to validate his/her code or make sure that a third-party solution is consistent with what is proposed by the community.

Although snippet matching is a difficult task, we believe that our methodology is a step toward the right direction. Future research includes further adapting the methodology to the special characteristics of different snippets, e.g., by clustering them according to their size or their complexity. Additionally, different structural representations for source code snippets may be tested in order to improve on the similarity scheme.

References

1. Diamantopoulos T, Symeonidis A (2015) Employing source code information to improve question-answering in stack overflow. In: Proceedings of the 12th Working Conference on Mining Software Repositories, MSR '15. IEEE Press, Piscataway, NJ, USA, pp 454–457
2. Ying ATT (2015) Mining challenge 2015: comparing and combining different information sources on the stack overflow data set. In: The 12th Working Conference on Mining Software Repositories, page to appear

3. Elasticsearch: RESTful, Distributed Search & Analytics. https://www.elastic.co/products/elasticsearch, 2016. [Retrieved: April 2016]

4. Holmes R, Murphy GC (2005) Using structural context to recommend source code examples. In: Proceedings of the 27th International Conference on Software Engineering, ICSE '05. ACM, New York, NY, USA, pp 117–125

5. Thummalapenta S, Xie T (2007) PARSEWeb: A Programmer Assistant for Reusing Open Source Code on the Web. In: Proceedings of the 22nd IEEE/ACM International Conference on Automated Software Engineering, ASE '07. ACM, New York, NY, USA, pp 204–213

6. Sahavechaphan N, Claypool K (2006) XSnippet: mining for sample code. SIGPLAN Not 41(10):413–430

7. Subramanian S, Holmes R (2013) Making sense of online code snippets. In: Proceedings of the 2013 10th IEEE Working Conference on Mining Software Repositories (MSR). pp 85–88

8. Subramanian S, Inozemtseva L, Holmes R (2014) Live API documentation. In: Proceedings of the 36th International Conference on Software Engineering, ICSE 2014. ACM, New York, NY, USA,pp 643–652

9. Ponzanelli L, Bavota G, Di Penta M, Oliveto R, Lanza M (2014) Mining stackoverflow to turn the IDE into a self-confident programming prompter. In: Proceedings of the 11th Working Conference on Mining Software Repositories, MSR 2014. ACM, New York, NY, USA, pp 102–111

10. Hsu SK, Lin SJ (2011) MACs: mining API code snippets for code reuse. Expert Syst Appl 38(6):7291–7301

11. Cormen TH, Leiserson CE, Rivest RL, Stein C (2009) Introduction to algorithms, 3rd edn. The MIT Press, pp 390–396

Part IV
Quality Assessment

Chapter 9
Providing Reusability-Aware Recommendations

9.1 Overview

In Chaps. 5 and 6 we focused on the problem of component-level source code reuse and viewed the problem from a functional point of view. As already indicated in the related work analyzed in Chap. 6, most efforts apply match-making mechanisms to provide functionally adequate software components to the developer. Certain systems have also employed test cases to ensure that the desired functionality is fulfilled [1, 2], while others have even implemented simple source code transformations in order to integrate the components directly into the source code of the developer [3].

Although these systems cover the functional criteria posed by the developer, they do not offer any assurance concerning the *reusability* of the source code. Reusing retrieved code can be a risky practice, considering no quality expert has assessed it. A possible solution is to exploit the power of the developers' corpus, as performed by Bing Developer Assistant [3], which promotes components that have been chosen by other developers. However, crowdsourcing solutions are not always accurate, considering that developers may require specialized components with specific quality criteria. Code Conjurer [1] works in that direction by selecting the less complex out of all functionally equivalent components, determined by the lines of code. However, defining the criteria for the complexity and generally for the reusability of a component is not trivial.

In this chapter, we present an RSSE that covers both the functional and the reusability aspects of components. Our system is called QualBoa [4] since it harnesses the power of Boa [5], an indexing service, to locate useful software components and compute quality metrics. Upon downloading the components from GitHub, QualBoa generates for each component both a functional score and a reusability index based on quality metrics. Furthermore, the generated reusability index is configurable to allow the involvement of quality experts.

© Springer Nature Switzerland AG 2020
T. Diamantopoulos and A. L. Symeonidis, *Mining Software Engineering Data for Software Reuse*, Advanced Information and Knowledge Processing,
https://doi.org/10.1007/978-3-030-30106-4_9

9.2 QualBoa: Reusability-Aware Component Recommendations

9.2.1 High-Level Overview of QualBoa

The architecture of QualBoa is shown in Fig. 9.1. This architecture is agnostic, since it can support any language and integrate with various code hosting services. We employ Boa and GitHub as the code hosting services, and implement the language-specific components for the Java programming language.

Initially, the developer provides a query in the form of a *signature* (step 1). As in Mantissa (see Chap. 6), a signature is similar to a Java interface and comprises all methods of the desired class component. An example signature for a "Stack" component is shown in Fig. 9.2.

The signature is given as input to the *Parser*. The Parser, which is implemented using Eclipse JDT,[1] extracts the *Abstract Syntax Tree (AST)* (step 2). Upon extracting the AST for the query, the *Boa Downloader* parses it and constructs a Boa query for relevant components, including also calculations for quality metrics. The query is submitted to the Boa engine [5] (step 3), and the response is a list of paths to Java files and the values of quality metrics for these files. The result list is given to the *GitHub Downloader* (step 4), which downloads the files from GitHub (step 5). The files are then given to the Parser (step 6), so that their ASTs are extracted (step 7).

The ASTs of the downloaded files are further processed by the *Functional Scorer*, which creates the *functional score* for each result (step 8). The Functional Score of a file denotes whether it fulfills the functionality posed by the query. The *Reusability*

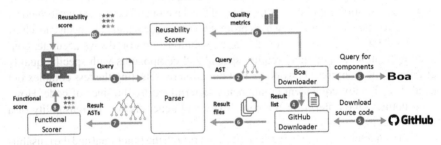

Fig. 9.1 The architecture of QualBoa

Fig. 9.2 Example signature for class "Stack" with methods "push" and "pop"

```
public class Stack{
    public void push(Object element);
    public Object pop();
}
```

---

[1] http://www.eclipse.org/jdt/.

Scorer receives as input the metrics for the files (step 9) and generates a *reusability score* (step 10), denoting the extent to which each file is reusable. Finally, the developer (*Client*) is provided with the scores for all files, and the results are ranked according to their Functional Score.

9.2.2 Downloading Source Code and Metrics

Upon extracting the elements of the query, i.e., class name, method names, and types, QualBoa finds useful software components and computes their metrics using Boa [5]. A simplified version of the query is shown in Fig. 9.3.

The query follows the visitor pattern to traverse the ASTs of the Java files in Boa. The first part of the query involves visiting the type declarations of all files and determining whether they match the given class name (class_name). The methods of any matched files are further visited to check whether their names and types match those of the query (method_names and method_types). The number of matched elements is aggregated to rank the results and a list with the highest ranked 150 results is given to the GitHub Downloader, which retrieves the Java files from the GitHub API.

The second part of the query involves computing source code metrics for the retrieved files. Figure 9.3 depicts the code for computing the number of public fields. We extend the query to compute the metrics shown in Table 9.1 (for which the rationale is provided in Sect. 9.2.4).

Table 9.1 The reusability metrics of QualBoa

Metric	Description
Average Lines of Code per Method	Average number of lines of code for each method
Average Cyclomatic Complexity	Average McCabe complexity for all methods
Coupling Between Objects	Number of classes that the class under analysis depends on
Lack of Cohesion in Methods	Number of method pairs of the class with common variables among them subtracted by the number of method pairs without common variables among them
Average Block Depth	Average control flow depth for all methods
Efferent Couplings	Number of data types of the class
Number of Public Fields	Total number of public fields of the class
Number of Public Methods	Total number of public methods of the class

```
visit(p, visitor {
    ...
    before node: Declaration -> {
        if (match(class_name, node.name)) {
            foreach (i: int; node.methods[i])
                visit(node.methods[i]);
        }
    }
    before node: Method -> {
        for (i := 0; i < len(method_names); i++) {
            if (match(method_names[i], node.name)) {
                match_names[i] = true;
                if (method_types[i] == node.return_type.name)
                    match_types[i] = true;
            }
        }
    }
    ...
    after node: Declaration -> {
        foreach (i: int; def(node.fields[i])) {
            foreach (j: int; def(node.fields[i].modifiers[j])) {
                if (node.fields[i].modifiers[j].visibility == Visibility.PUBLIC)
                    num_public_fields++;
            }
        }
        ...
    }
});
```

Fig. 9.3 Boa query for components and metrics

9.2.3 Mining Source Code Components

The Functional Scorer computes the similarity between each of the ASTs of the results with the AST of the query, in order to rank the results according to the functional desiderata of the developer. Initially, both the query and each examined result file is represented as a list. The list of elements for a result file is defined in the following equation:

$$result = [name, method_1, method_2, \ldots, method_n] \qquad (9.1)$$

where *name* is the name of the class and $method_i$ is a sublist of the ith method of the class, out of n methods in total. The sublist of a method is defined as follows:

$$method = [name, type, param_1, param_2, \ldots, param_m] \tag{9.2}$$

where *name* is the name of the method, *type* is its return type, and $param_j$ is the type of its jth parameter, out of m parameters in total. Using Eqs. (9.1) and (9.2), we can represent each result, as well as the query, as a nested list structure.

Comparing a query to a result requires computing a similarity score between the two corresponding lists, which in turn implies computing the score between each pair of methods. Note that computing the maximum similarity score between two lists of methods requires finding the score of each individual pair of methods and selecting the highest scoring pairs. Following the approach of the *stable marriage problem* (see Chap. 6), this is accomplished by ordering the pairs according to their similarity and selecting them one-by-one off the top, noticing whether a method has already been matched. The same procedure is also applied for optimally matching the parameters between the two method lists.

Class names, method names, and types, as well as parameter types are matched using a token set similarity approach. Since in Java identifiers follow the camelCase convention, we first split each string into tokens and compute the Jaccard index for the two token sets. Given two sets, the Jaccard index is defined as the size of their intersection divided by the size of their union. Finally, the similarity between two lists or vectors **A** and **B** (either in method or in class/result level) is computed using the Tanimoto coefficient [6] of the vectors $\mathbf{A} \cdot \mathbf{B} / (|\mathbf{A}|^2 + |\mathbf{B}|^2 - \mathbf{A} \cdot \mathbf{B})$, where $|\mathbf{A}|$ and $|\mathbf{B}|$ are the sizes of vectors **A** and **B**, and $\mathbf{A} \cdot \mathbf{B}$ is their inner product.

For instance, given the query list $[Stack, [push, void, Object], [pop, Object]]$ and the component list $[IntStack, [pushObject, void, int], [popObject, int]]$, the similarity between the method pairs for the "push" functionality is computed as follows:

$$
\begin{aligned}
score_{PUSH} = Tanimoto([1, 1, 1], & \tag{9.3} \\
[Jaccard(\{push\}, \{push, object\}), & \\
Jaccard(\{void\}, \{void\}) & \\
Jaccard(\{object\}, \{int\})]) & \\
= Tanimoto([1, 1, 1], [0.5, 1, 0]) \simeq 0.545 &
\end{aligned}
$$

where the query list is always a list with all elements set to 1, since it constitutes a perfect match. Similarly, the score for the "pop" functionality is computed as follows:

$$score_{POP} = Tanimoto([1, 1], \qquad\qquad\qquad\qquad (9.4)$$
$$[Jaccard(\{pop\}, \{pop, object\}),$$
$$Jaccard(\{object\}, \{int\})])$$
$$= Tanimoto([1, 1], [0.5, 0]) \simeq 0.286$$

Finally, the total score for the class is computed as follows:

$$score_{STACK} = Tanimoto([1, 1, 1], \qquad\qquad\qquad\qquad (9.5)$$
$$[Jaccard(\{stack\}, \{int, stack\}),$$
$$score_{PUSH},$$
$$score_{POP})$$
$$= Tanimoto([1, 1, 1], [0.5, 0.545, 0.286]) \simeq 0.579$$

9.2.4 Recommending Quality Code

Upon constructing a functional score for the retrieved components, QualBoa checks whether each component is suitable for reuse using source code metrics. Although the problem of measuring the reusability of a component using source code metrics has been studied extensively [7–10], the choice of metrics is not trivial; research efforts include employing the C&K metrics [8], coupling and cohesion metrics [9], and several other metrics referring to volume and complexity [7]. In any case, however, the main quality axes that measure whether a component is reusable are common. In accordance with the definitions of different quality characteristics [11] and the current state of the art [8, 10, 12], reusability spans across the *modularity*, *usability*, *maintainability*, and *understandability* concepts.

Our reusability model, shown in Table 9.2, includes eight metrics that refer to one or more of the aforementioned quality characteristics. These metrics cover several aspects of a source code component, including volume, complexity, coupling, and cohesion. Apart from covering the above criteria, the eight metrics were selected as they can be computed by the Boa infrastructure. The reader is referred to the following chapter for a more detailed model on reusability. This model is fairly simple, marking the value of each metric as *normal* or *extreme* according to the thresholds shown in the second column of Table 9.2. When a value is flagged as normal, then it contributes to one quality point in the relevant characteristics. Reusability includes all eight values. Thus, given, e.g., a component with extreme values for 2 out of 8 metrics, the reusability score would be 6 out of 8.

Determining appropriate thresholds for quality metrics is non-trivial, since different types of software may have to reach specific quality objectives. Thus, QualBoa offers the ability to configure these thresholds, while their default values are set

Table 9.2 The reusability model of QualBoa

Quality metrics	Extremes	Quality characteristics				
		Modularity	Maintainability	Usability	Understandability	Reusability
Average lines of code per method	>30		×		×	×
Average cyclomatic complexity	>8		×		×	×
Coupling between Objects	>20	×		×		×
Lack of cohesion in methods	>20	×				×
Average block depth	>3				×	×
Efferent couplings	>20	×		×		×
Number of public fields	>10			×		×
Number of public methods	>30		×	×		×
#Metrics per quality characteristic:		3	4	4	3	8

Table 9.3 Dataset for the evaluation of QualBoa

Class	Methods
Calculator	add, sub, div, mult
ComplexNumber	ComplexNumber, add, getRealPart, getImaginaryPart
Matrix	Matrix, set, get, multiply
MortgageCalculator	setRate, setPrincipal, setYears, getMonthlyPayment
ShoppingCart	getItemCount, getBalance, addItem, empty, removeItem
Spreadsheet	put, get
Stack	push, pop

according to the current state of the practice, as defined by current research [13] and widely used static code analysis tools, such as PMD[2] and CodePro AnalytiX.[3]

9.3   Evaluation

We evaluated QualBoa in the query dataset of Table 9.3, which contains seven queries for different types of components. This dataset is similar to the one presented in Chap. 6 for the evaluation of Mantissa against Code Conjurer [1]. Note, however, that this time the backend of our RSSE is Boa so the queries have to be adjusted in order to retrieve useful results. All queries returned results from Boa, except MortgageCalculator, for which we adjusted the class name to Mortgage, and the query filters to account only for the setRate method. Note, however, that the Functional Scorer of QualBoa uses all methods.

For each query, we examine the first 30 results of QualBoa, and mark each result as useful or not useful, considering a result useful if integrating it in the developer's code would require minimal or no effort. In other words, useful components are the ones that are understandable and at the same time cover the required functionality. Benchmarking and annotation were performed separately to ensure minimal threats to validity.

Given these marked results, we check the number of relevant results retrieved for each query, and compute also the average precision for the result rankings. The average precision was considered as the most appropriate metric since it covers not only the relevance of the results but also, more importantly, their ranking. We further assess the files using the reusability model of QualBoa and, upon normalizing to percentages, report the average reusability score for the useful/relevant results

---

[2]https://pmd.github.io/.

[3]https://developers.google.com/java-dev-tools/codepro/.

Table 9.4 Evaluation results of QualBoa

Query	#Relevant results	Average precision%	Reusability score%
Calculator	18	59.27	70.83
ComplexNumber	15	86.18	82.76
Matrix	10	95.88	88.68
MortgageCalculator	7	100.00	87.17
ShoppingCart	13	100.00	100.00
Spreadsheet	2	100.00	88.54
Stack	22	77.59	100.00
Average	12.43	88.42	88.28

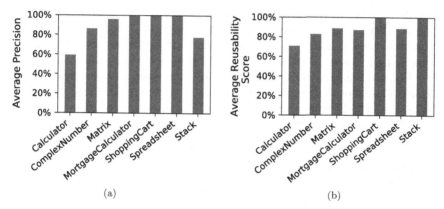

Fig. 9.4 Evaluation results of QualBoa depicting **a** the average precision and **b** the average reusability score for each query

of each query. The results of our evaluation are summarized in Table 9.4, while Fig. 9.4 graphically depicts the average precision value for each query and the average reusability score for each query.

Concerning the number of relevant results, QualBoa seems to be quite effective. In specific, our RSSE successfully retrieves at least 10 useful results for 5 out of 7 queries. Additionally, almost all queries have average precision scores higher than 75%, indicating that the results are also ranked correctly. This is particularly obvious in cases where the number of retrieved results is limited, such as the Mortgage-Calculator or the Spreadsheet component. The perfect average precision scores indicate that the relevant results for these queries are placed on the top of the ranking.

The results are also highly reusable, as indicated by the reusability score for each query. In 5 out of 7 queries, the average reusability score of the relevant results is quite near or higher than 87.5%, indicating that on average the components do not surpass more than 1 of the thresholds defined in Table 9.2, so they have a reusability score of at least 7 out of 8. The reusability score is also related to the overall complexity of the queried component. In specific, data structure components such as Stack or ShoppingCart are not prone to severe quality issues and thus have perfect scores. On the other hand, complex components such as Calculator or ComplexNumber may contain methods that interoperate, thus their reusability scores are expected to be lower.

9.4   Conclusion

In this chapter, we presented QualBoa, a system that incorporates functional and quality information to recommend components that are not only functionally equivalent to the query of the developer but also reusable. Our evaluation indicates that QualBoa is effective for retrieving reusable results. Further work lies in several directions. The main axis, however, is the reusability model of QualBoa. The current model is rather simple and was used mainly for validating the concept of this chapter, while, in the following chapter, we propose a strategy for the automated construction of a more effective reusability model. Finally, as a future idea, we may also note that the reusability model could be automatically adapted so that the recommended components have the same quality characteristics as the source code of the developer.

References

1. Hummel O, Janjic W, Atkinson C (2008) Code conjurer: pulling reusable software out of thin air. IEEE Softw 25(5):45–52
2. Reiss SP (2009) Semantics-based code search. In: Proceedings of the 31st International Conference on Software Engineering, ICSE '09. IEEE Computer Society, Washington, DC, USA, pp 243–253
3. Wei Y, Chandrasekaran N, Gulwani S, Hamadi Y (2015) Building Bing Developer Assistant. Technical Report MSR-TR-2015-36, Microsoft Research
4. Diamantopoulos T, Thomopoulos K, Symeonidis A (2016) QualBoa: reusability-aware Recommendations of source code components. In: Proceedings of the IEEE/ACM 13th Working Conference on Mining Software Repositories, MSR '16. pp 488–491
5. Dyer R, Nguyen HA, Rajan H, Nguyen TN (2013) Boa: a language and infrastructure for analyzing ultra-large-scale software repositories. In: 35th International Conference on Software Engineering, ICSE 2013. pp 422–431
6. Tanimoto TT (1957) IBM Internal Report
7. Caldiera G, Basili VR (1991) Identifying and qualifying reusable software components. Computer 24(2):61–70
8. Moser R, Sillitti A, Abrahamsson P, Succi G (2006) Does refactoring improve reusability? In: Proceedings of the 9th International Conference on Reuse of Off-the-Shelf Components, ICSR'06. Berlin, Heidelberg, Springer-Verlag, pp 287–297
9. Gui G, Scott PD (2006) Coupling and cohesion measures for evaluation of component reusability. In: Proceedings of the 2006 International Workshop on Mining Software Repositories, MSR '06. ACM, New York, NY, USA, pp 18–21
10. Poulin JS (1994) Measuring software reusability. In: Proceedings of the Third International Conference on Software Reuse: Advances in Software Reusability. pp 126–138

11. Spinellis D (2006) Code Quality: The Open Source Perspective. Effective software development series. Addison-Wesley Professional
12. Dandashi F (2002) A method for assessing the reusability of object-oriented code using a validated set of automated measurements. In: Proceedings of the 2002 ACM Symposium on Applied Computing, SAC '02. ACM, New York, NY, USA, pp 997–1003
13. Ferreira KAM, Bigonha MAS, Bigonha RS, Mendes LFO, Almeida HC (2012) Identifying thresholds for object-oriented software metrics. J Syst Softw 85(2):244–257

Chapter 10
Assessing the Reusability of Source Code Components

10.1 Overview

As already mentioned, the quality assessment of software components is very important in the context of source code reuse. The process of integrating components in one's own source code may lead to failures, since the reused code may not satisfy basic functional or non-functional requirements. Thus, the quality assessment of components to be reused poses a major challenge for the research community.

An important aspect of this challenge is the fact that quality is context-dependent and may mean different things to different people [1]. Therefore, the need for standardization arises to create a common reference. Toward this direction, the international standards ISO/IEC 25010:2011 [2] and ISO/IEC 9126 [3] define a quality model that consists of eight major quality attributes and several quality properties. According to these standards, software reusability, i.e., the extent to which a software component is reusable, is related to four major quality properties: *Understandability*, *Learnability*, *Operability*, and *Attractiveness*. These properties are directly associated with *Usability*, and further affect *Functional Suitability*, *Maintainability* and *Portability*, thus covering all four quality attributes related to reusability [4, 5] (the rest characteristics are *Reliability*, *Performance and Efficiency*, *Security*, and *Compatibility*).

Current research efforts largely focus on quality attributes such as maintainability or security, and assessing reusability has not yet been extensively addressed. There are several approaches that aspire to assess the quality [4–7] and the reusability [8, 9] of source code components using static analysis metrics, such as the known CK metrics [10]. These approaches, however, are based on metric thresholds that, whether defined manually [4, 6, 7] or derived automatically using a model [11–14], are usually constrained by the lack of objective ground truth values for software quality. As a result, they typically resort to expert help, which may be subjective, case-specific, or even unavailable [15, 16].

© Springer Nature Switzerland AG 2020
T. Diamantopoulos and A. L. Symeonidis, *Mining Software Engineering Data for Software Reuse*, Advanced Information and Knowledge Processing, https://doi.org/10.1007/978-3-030-30106-4_10

In this chapter, we mitigate the need for such an imposed ground truth quality value, by employing user-perceived quality as a measure of the reusability of a software component. From a developer perspective, we are able to associate user-perceived quality with the extent to which a software component is adopted by developers, thus arguing that the popularity of a component can be a measure of its reusability. We provide a proof-of-concept for this hypothesis and further investigate the relationship between popularity and user-perceived quality by using a dataset of source code components along with their popularity as determined by their rating on GitHub. Based on this argument, we design and implement a methodology that covers the four software quality properties related to reusability by using a large set of static analysis metrics. Given these metrics and using the popularity of GitHub projects as ground truth, we can estimate the reusability of a source code component. In specific, we model the behavior of the metrics to automatically translate their values into a reusability score [17]. Metric behaviors are then used to train a set of reusability estimation models using different state-of-the-practice machine learning algorithms. Those models are able to estimate the reusability degree of components at both class and package level, as perceived by software developers.

10.2 Background on Reusability Assessment

Most of the existing quality and reusability assessment approaches make use of static analysis metrics in order to train quality estimation models [6, 8, 9, 18]. In principle, estimating quality through static analysis metrics is a non-trivial task, as it often requires determining quality thresholds [4], which is usually performed by experts who manually examine the source code [19]. However, the manual examination of source code, especially for large complex projects that change on a regular basis, is not always feasible due to constraints in time and resources. Moreover, expert help may be subjective and highly context-specific.

Other approaches may require multiple parameters for constructing quality evaluation models [15], which are again highly dependent on the scope of the source code and are easily affected by subjective judgment. Thus, a common practice involves deriving metric thresholds by applying machine learning techniques on a benchmark repository. Ferreira et al. [20] propose a methodology for estimating thresholds by fitting the values of metrics into probability distributions, while Alves et al. [21] follow a weight-based approach to derive thresholds by applying statistical analysis on the metrics values. Other approaches involve deriving thresholds using bootstrapping [22] and ROC curve analysis [23]. Still, these approaches are subject to the projects selected for the benchmark repository. Finally, there are some approaches that focus specifically on reusability [12–14, 24]. As in quality estimation, these approaches also first involve quantifying reusability through reuse-related information such as reuse frequency [24]. Then, machine learning techniques are employed to train reusability evaluation models using as input the values of static analysis metrics [12–14].

Although the aforementioned approaches can be effective for certain cases, their applicability in real-world scenarios is limited. At first, approaches using predefined thresholds [4, 6, 7, 19], even with the guidance of an expert, cannot extend their scope beyond the specific class and the characteristics of the software for which they have been designed. Automated quality and reusability evaluation systems appear to overcome these issues [11–14, 20–23]. However, they are also still confined by the ground truth knowledge of an expert for training the model, and thus do not rate the reusability of source code according to the diverse needs of the developers. As a result, several academic and commercial approaches [15, 25] resort to providing detailed reports of quality metrics, leaving configurations to developers or experts.

To overcome these issues, we build a reusability estimation system to provide a single score for every class and every package of a given software component. Our system does not use predefined metric thresholds, while it also refrains from expert representations of quality; instead, we use popularity (i.e., user-perceived quality) as ground truth for reusability [26]. Given that popular components have a high rate of reuse [27], by intuition the preference of developers may be a measure of high reusability. We employ this concept to evaluate the impact of each metric into the reusability degree of software components individually, and subsequently aggregate the outcome of our analysis in order to construct a final reusability score. Finally, we quantify the reusability degree of software components both at class and package level by training reusability estimation models that effectively estimate the degree to which a component is reusable as perceived by software developers.

10.3 Correlating Reusability with Popularity

In this section, we discuss the potential connection between reusability and popularity and indicate the compliance of our methodology with the ISO quality standards.

10.3.1 GitHub Popularity as Reusability Indicator

To support our claim that reusability is correlated with popularity, we computed the correlation between the number of stars and the number of forks for the 100 most popular Java repositories of GitHub (i.e., the ones with the most stars) [26]. As illustrated in Fig. 10.1, there is a strong positive correlation between the two metrics. In specific, the value of the correlation coefficient is 0.68.

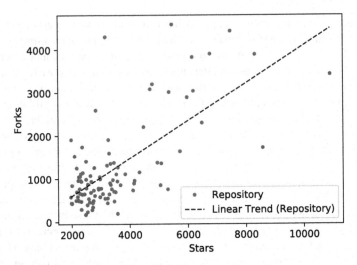

Fig. 10.1 Diagram depicting the number of stars versus the number of forks for the 100 most popular GitHub repositories

Table 10.1 Categories of static analysis metrics related to reusability

Categories	Forks-related			Stars-related
	Understandability	Learnability	Operability	Attractiveness
Complexity	IP	IP	IP	P
Coupling	IP	IP	IP	P
Cohesion	P	P	P	P
Documentation	P	P	–	–
Inheritance	IP	IP	–	P
Size	IP	IP	IP	P

P: Proportional Relation
IP: Inverse-Proportional Relation

10.3.2 Modeling GitHub Popularity

We associate reusability with the following major properties described in ISO/IEC 25010 [2] (and 9126 [3]): *Understandability, Learnability, Operability*, and *Attractiveness*. According to research performed by ARiSA [28], various static analysis metrics are highly related to these properties. Table 10.1 summarizes the relations between the six main categories of static analysis metrics and the aforementioned reusability-related quality properties. "P" is used to show proportional relation and "IP" implies inverse-proportional relation.

As already mentioned, we employ GitHub stars and forks in order to quantify reusability and subsequently associate it with the aforementioned properties. Since forks measure how many times the software repository has been cloned, they can be associated with *Understandability*, *Learnability*, and *Operability*, as these properties formulate the degree to which a software component is reusable. Stars, on the other hand, reflect the number of developers that found the repository interesting (and decide to follow its updates), thus we may use them as a measure of its *Attractiveness*.

10.4   Reusability Modeling

In this section, we present an analysis of reusability from quality attributes perspective and design a reusability score using GitHub information and static analysis metrics.

10.4.1   Benchmark Dataset

We created a dataset that includes the values of the static analysis metrics shown in Table 10.2, for the 100 most starred and 100 most forked GitHub Java projects (137 in total). These projects amount to more than 12M lines of code spread in almost 15 K packages and 150 K classes. All metrics were extracted at class and package level using SourceMeter [29].

10.4.2   Evaluation of Metrics' Influence on Reusability

As GitHub stars and forks refer to repository level, they are not adequate on their own for estimating the reusability of class level and package level components. Thus, we estimate reusability using static analysis metrics. For each metric, we first perform distribution-based binning and then relate its values to those of the stars/forks to incorporate reusability information. The final reusability estimation is computed by aggregating over the estimations derived by each metric.

Distribution-Based Binning
As the values of each static analysis metric are distributed differently among the repositories of our dataset, we first define a set of representative intervals (bins), unified across repositories, that effectively approximate the actual distribution of the values. For this purpose, we use the values of each metric from all the packages (or classes) of our dataset to formulate a generic distribution, and then determine the optimal bin size that results in the minimum information loss. In an effort to follow a generic and accurate binning strategy while taking into consideration the fact that the values of the metrics do not follow a normal distribution, we use the *Doane* formula [30] for selecting the bin size in order to account for the skewness of the data.

Figure 10.2 depicts the histogram of the API Documentation (AD) metric at package level, which will be used as an example throughout this section. Following our

Table 10.2 Overview of static metrics and their applicability on different levels

Name	Description	Class	Package
Complexity			
NL {•,E}	Nesting Level {Else-If}	×	
WMC	Weighted Methods per Class	×	
Coupling			
CBO {•,I}	Coupling Between Object classes {Inverse}	×	
N {I,O}I	Number of { Incoming, Outgoing} Invocations	×	
RFC	Response set For Class	×	
Cohesion			
LCOM5	Lack of Cohesion in Methods 5	×	
Documentation			
AD	API Documentation	×	
{•,T}CD	{Total} Comment Density	×	×
{•,T}CLOC	{Total} Comment Lines of Code	×	×
DLOC	Documentation Lines of Code	×	
P {D,U}A	Public {Documented, Undocumented} API	×	×
TAD	Total API Documentation		×
TP {D,U}A	Total Public {Documented, Undocumented} API		×
Inheritance			
DIT	Depth of Inheritance Tree	×	
NO {A,C}	Number of {Ancestors, Children}	×	
NO {D,P}	Number of {Descendants, Parents}	×	
Size			
{•,T} {•,L}LOC	{Total} {Logical} Lines of Code	×	×
N{G,S}	Number of {Getters, Setters}	×	×
N{A,M}	Number of {Attributes, Methods}	×	×
N{CL,EN,IN,P}	Number of {Classes, Enums, Interfaces, Packages}		×
NL{G,S}	Number of Local {Getters, Setters}	×	
NL{A,M}	Number of Local {Attributes, Methods}	×	
NLP{A,M}	Number of Local Public {Attributes, Methods}	×	
NP{A,M}	Number of Public {Attributes, Methods}	×	×
NOS	Number of Statements	×	
TNP{CL,EN,IN}	Total Number of Public {Classes, Enums, Interfaces}		×
TN{CL,DI,EN,FI}	Total Number of {Classes, Directories, Enums, Files}		×

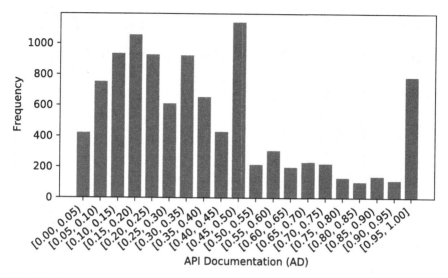

Fig. 10.2 Package-level distribution of API Documentation for all repositories

binning strategy, 20 bins are produced. AD values appear to have positive skewness, while their highest frequency is in the interval [0.1, 0.5]. After having selected the appropriate bins for each metric, we construct the histograms for each of the 137 repositories.

Figure 10.3 depicts the histograms for two different repositories, where it is clear that each repository follows a different distribution for the values of AD. This is expected, as every repository is an independent entity with a different scope, developed by different contributors; thus, it accumulates a set of different characteristics.

Relating Bins Values with GitHub Stars and Forks

We use the produced histograms (one for each repository using the bins calculated in the previous step) in order to construct a set of data instances that relate each metric bin value (here the AD) to a GitHub stars (or forks) value. So, we aggregate these values for all the bins of each metric, i.e., for metric X and bin 1 we gather all stars (or forks) values that correspond to packages (or classes) for which the metric value lies in bin 1 and aggregate them using an averaging method. This process is repeated for every metric and every bin.

Practically, for each bin value of the metric, we gather all relevant data instances and calculate the weighted average of their stars (or forks) count, which represents the stars-based (or forks-based) reusability value for the specific bin. The reusability scores are defined using the following equations:

$$RS_{Metric}(i) = \sum_{repo=1}^{N} freq_{p.u.}(i) \cdot \log(S(repo)) \qquad (10.1)$$

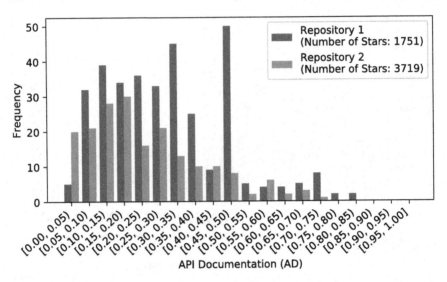

Fig. 10.3 Package-level distribution of API Documentation for two repositories

$$RF_{Metric}(i) = \sum_{repo=1}^{N} freq_{p.u.}(i) \cdot \log(F(repo)) \qquad (10.2)$$

where $RS_{Metric}(i)$ refers to the stars-based reusability score of the i-th bin for the metric under evaluation and $RF_{Metric}(i)$ refers to the respective forks-based reusability score. $S(repo)$ and $F(repo)$ refer to the number of stars and the number of forks of the repository, respectively, while the use of logarithm acts as a smoothing factor between the big differences in the number of stars and forks among the repositories. Finally, the term $freq_{p.u.}(i)$ is the normalized/relative frequency of the metric value of the i-th bin, defined as follows:

$$freq_{p.u.}(i) = \frac{F_i}{\sum_{i=1}^{N} F_i} \qquad (10.3)$$

where F_i is the absolute frequency (i.e., the count) of the metric values lying in the i-th bin. For example, if a repository had 3 bins with 5 AD values in bin 1, 8 values in bin 2, and 7 values in bin 3, then the normalized frequency for bin 1 would be $5/(5+8+7) = 0.25$, for bin 2 it would be $8/(5+8+7) = 0.4$, and for bin 3 it would be $7/(5+8+7) = 0.35$. The use of normalized frequency was chosen to eliminate any biases caused by the high variance in the number of packages among the different repositories. Given that the selected repositories are the top 100 starred and forked repositories, these differences do not necessarily reflect the actual differences in the reusability degree of packages.

Figure 10.4 illustrates the results of applying Eq. (10.1), i.e., the star-based reusability, to the AD metric values at package level. Note also that the upper and

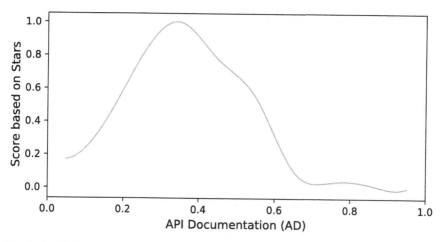

Fig. 10.4 API Documentation versus star-based reusability score

lower 5% of the metric values are removed at this point as outliers. As shown in this figure, the reusability score based on AD is maximum for AD values in the interval [0.25, 0.5].

Finally, based on the fact that we compute a forks-based and a stars-based reusability score for each metric, the final reusability score for each source code component (class or package) is given by the following equations:

$$RS_{Final} = \frac{\sum_{j=1}^{k} RS_{metric}(j) \cdot corr(metric_j, stars)}{\sum_{j=1}^{k} corr(metric_j, stars)} \quad (10.4)$$

$$RF_{Final} = \frac{\sum_{j=1}^{k} RF_{metric}(j) \cdot corr(metric_j, forks)}{\sum_{j=1}^{k} corr(metric_j, forks)} \quad (10.5)$$

$$Final_{Score} = \frac{3 \cdot RF_{Final} + RS_{Final}}{4} \quad (10.6)$$

where k is the number of metrics at each level. RS_{Final} and RF_{Final} correspond to the final stars-based and forks-based scores, respectively. $RS_{metric}(j)$ and $RF_{metric}(j)$ refer to the scores for the j-th *metric* as given by Eqs. (10.1) and (10.2), while $corr(metric_j, stars)$ and $corr(metric_j, forks)$ correspond to the Pearson correlation coefficient between the values of j-th *metric* and the stars or forks, respectively. $Final_{Score}$ refers to the final reusability score and is the weighted average of the stars-based and forks-based scores. More weight (3 vs 1) is given in the forks-based score as it is associated with more reusability-related quality attributes (see Table 10.1).

10.5 Reusability Estimation

This section presents our proposed methodology for software reusability estimation and illustrates the construction of our models.

10.5.1 System Overview

The data flow of our system is shown in Fig. 10.5. The input is the set of static analysis metrics, along with the popularity information (stars and forks) extracted from GitHub repositories. As the system involves reusability estimation both at class and package levels, we train one model for each individual metric applied at each level using the results of the aforementioned analysis. The output of each model provides a reusability score that originates from the value of the class (or package) under evaluation for the respective metric. All the scores are then aggregated to calculate a final reusability score that represents the degree to which the class (or package) is adopted by developers.

10.5.2 Model Construction

In an effort to construct effective models, each fitting the behavior of one metric, we perform a comparative analysis between different state-of-the-practice machine learning algorithms. In specific, we evaluate three techniques: *Support Vector Regression*, *Polynomial Regression*, and *Random Forest Regression*. The Random Forest models have been constructed using the bagging ensemble/resampling method, while for the Support Vector Regression models we used a Radial Basis Function (RBF) kernel. Regarding the polynomial regression models, the fitting procedure involves choosing the optimal degree by applying the elbow method on the square sum of

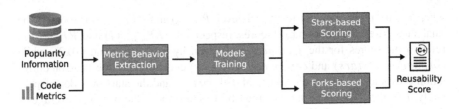

Fig. 10.5 Overview of the reusability evaluation system

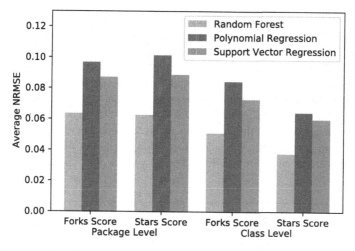

Fig. 10.6 Average NRMSE for all three machine learning algorithms

residuals.[1] To account for cases where the number of bins, and consequently the number of training data points, is low, we used linear interpolation up to the point where the dataset for each model contained 60 instances.

To evaluate the effectiveness of our constructed models and select the one that best matches the identified behaviors, we employ the *Normalized Root Mean Squared Error (NRMSE)* metric. Given the actual scores $y_1, y_2, \ldots, y_n$, the predicted scores $\hat{y}_1, \hat{y}_2, \ldots, \hat{y}_n$, and the mean actual score $\bar{y}$, the NRMSE is calculated as follows:

$$NRMSE = \sqrt{\frac{1}{N} \cdot \frac{\sum_{i=1}^{N} (\hat{y}_i - y_i)^2}{\sum_{i=1}^{N} (\bar{y} - y_i)^2}} \qquad (10.7)$$

where N is the number of samples in the dataset. NRMSE was selected as it does not only take into account the average difference between the actual and the predicted values, but also provides a comprehensible result in a certain scale.

Figure 10.6 presents the mean NRMSE for the three algorithms regarding the reusability scores (forks/stars based) at both class and package levels. Although all three approaches seem quite effective as their NRMSE values are low, the Random Forest clearly outperforms the other two algorithms in all four categories, while the Polynomial Regression models exhibit the highest errors. Another interesting conclusion is that reusability estimation at package level exhibits higher errors for all three approaches. This fact originates from our metrics behavior extraction methodology, which uses ground truth information at repository level. Given that the number of classes greatly outnumbers the number of packages within a given repository, the

---

[1]According to this method, the optimal degree is the one for which there is no significant decrease in the square sum of residuals when increasing the order by one.

extracted behaviors based on packages are less resilient to noise and thus result in higher errors. Finally, as the Random Forest is more efficient than the other two approaches, we select it as the main algorithm for the rest of the evaluation.

Figures 10.7a and 10.7b depict the distribution of the reusability score at class and package levels, respectively. As expected, the score in both cases follows a distribution similar to normal and the majority of instances are accumulated evenly around 0.5. For the reusability score at class level, we observe a left-sided skewness. After manual inspection of the classes that received scores in the interval [0.2, 0.25], they appear to contain little valuable information (e.g., most of them have LOC < 10) and thus are given low reusability score.

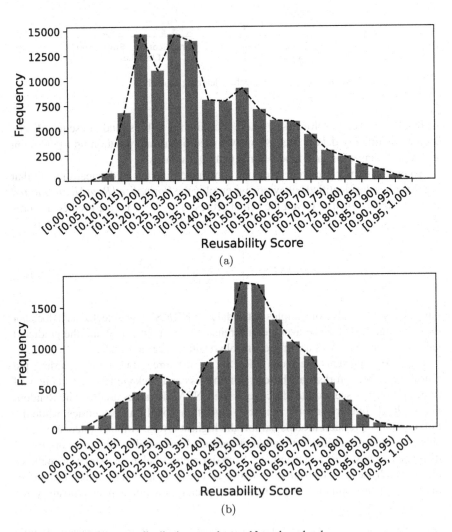

Fig. 10.7 Reusability score distribution at **a** class and **b** package level

Table 10.3 Dataset statistics

Project/Repository name	Type	#Packages	#Classes
realm/realm-java	Human-generated	137	3859
liferay/liferay-portal	Human-generated	1670	3941
spring-projects/spring-security	Human-generated	543	7099
Webmarks	Auto-generated	9	27
WSAT	Auto-generated	20	127

10.6 Evaluation

Both human-generated and auto-generated projects were selected to evaluate our models under different scenarios. In specific, we used the five projects shown in Table 10.3, out of which three were retrieved from GitHub and two were automatically generated using the tools of S-CASE.[2] Comparing the output of the system on these two types of projects shall provide interesting insight. Human-generated projects are expected to exhibit high deviations in their reusability score, as they include variable sets of components. By contrast, auto-generated projects are RESTful services and, given their automated structure generation, are expected to have high reusability scores and low deviations. Additionally, as also shown in Table 10.3, the projects have different characteristics in terms of size (number of classes and packages) and thus can be used to evaluate the capability of our reusability model to generalize to components from multiple sources.

In the following subsections, we provide an analysis of the reusability estimation of these projects as well as an estimation example for certain classes and packages.

10.6.1 Reusability Estimation Evaluation

Figures 10.8a and 10.8b depict the distributions of the reusability score for all projects at class level and package level, respectively. The boxplots in blue refer to the human-generated projects, while the ones in orange refer to the auto-generated ones.

At first, it is obvious that the variance of the reusability scores is higher in the human-generated projects than in the auto-generated ones. This is expected since the projects that were generated automatically have proper architecture and high abstraction levels. These two projects also have similar reusability values, which is due to the fact that projects generated from the same tool ought to share similar characteristics that are reflected in the values of the static analysis metrics.

---

[2]http://s-case.github.io/.

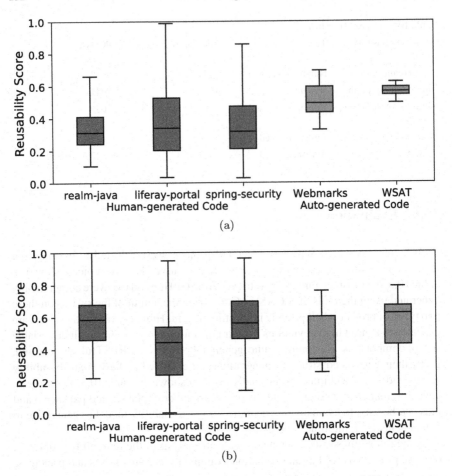

Fig. 10.8 Boxplots depicting reusability distributions for three human-generated (■) and two auto-generated (■) projects, **a** at class level and **b** at package level (color online)

The high variance for the human-generated projects indicates that our reusability assessment methodology is capable of distinguishing components with both high and low degrees of reusability. Typically, these projects involve several components, of which others were written while keeping in mind software reuse (e.g., APIs) and others not. Another effect reflected on these values is the number of distinct developers working on each project, a metric which is not an issue for the automatically generated projects. Finally, given the reusability score distribution for all projects, we conclude that the results are consistent regardless of the project size. Despite the fact that the number of classes and the number of packages vary from very low values (e.g., only 9 packages and 27 classes) to very high values (e.g., 1670 packages and 7099 classes), the score is not affected.

Table 10.4 Metrics for classes and packages with different reusability

Metrics	Classes		Packages	
	Class 1	Class 2	Package 1	Package 2
WMC	14	12	–	–
CBO	12	3	–	–
LCOM5	2	11	–	–
CD (%)	20.31%	10.2%	41.5%	0.0%
RFC	12	30	–	–
LOC	84	199	2435	38
TNCL	–	–	8	2
Reusability score	95.78%	10.8%	90.59%	16.75%

10.6.2 Example Reusability Estimation

Further assessing the validity of the reusability scores, we manually examined the static analysis metrics of sample classes and packages in order to check whether they align with the reusability estimation. Table 10.4 provides an overview of a subset of the computed metrics for representative examples of classes and packages with different reusability scores. The table contains static analysis metrics for two classes and two packages that received both high and low reusability scores.

Examining the values of the computed static analysis metrics, one can see that the reusability estimation in all four cases is reasonable and complies with the literature findings on the impact of metrics on the quality properties related to reusability (see Table 10.1). Concerning the class that received high reusability estimation, it appears to be well documented (the value of the Comments Density (CD) is 20.31%), which indicates that is suitable for reuse. In addition, the value of its Lines of Code (LOC) metric, which is highly correlated with the degree of understandability and thus the reusability, is typical. As a result, the score for this class is quite rational. On the other hand, the class that received low score appears to have low cohesion (the LCOM5 metric value is high) and high coupling (the RFC metric value is high). Those code properties are crucial for the reusability-related quality attributes and thus the low score is expected.

Finally, upon examining the static analysis metrics of the packages, the conclusions are similar. The package that received high reusability score appears to have proper documentation level (based on CD value) which combined with its typical size makes it suitable for reuse. This is reflected in the reusability score. On the other hand, the package that received low score appears to have no valuable information due to the fact that its size is very small (only 38 lines of code).

10.7 Conclusion

In this chapter, we proposed a source code quality estimation approach according to which user-perceived quality is employed as a measure of the reusability of a software component using code popularity as ground truth information. Thus, we formed a reusability score for both classes and packages using the GitHub stars and forks. Our evaluation indicates that our approach can be effective for estimating reusability at class and package level.

Concerning future work, an interesting idea would be to investigate and possibly redesign the reusability score in a domain-specific context. In addition, we plan to focus on creating models for different quality axes [31, 32]. Finally, evaluating our system under realistic reuse scenarios, possibly also involving software developers, would be useful to further validate our approach.

References

1. Pfleeger SL, Kitchenham B (1996) Software quality: the elusive target. IEEE Software, pp 12–21
2. ISO/IEC 25010:2011 (2011) https://www.iso.org/standard/35733.html. Retrieved: November 2017
3. ISO/IEC 9126-1:2001 (2001) https://www.iso.org/standard/22749.html. Retrieved: October 2017
4. Diamantopoulos T, Thomopoulos K, Symeonidis A (2016) Symeonidis. QualBoa: reusability-aware recommendations of source code components. In: Proceedings of the IEEE/ACM 13th Working Conference on Mining Software Repositories, MSR '16, pp 488–491
5. Taibi F (2014) Empirical analysis of the reusability of object-oriented program code in open-source software. Int J Comput Inf Syst Control Eng 8(1):114–120
6. Le Goues C, Weimer W (2012) Measuring code quality to improve specification mining. IEEE Trans Softw Eng 38(1):175–190
7. Washizaki H, Namiki R, Fukuoka T, Harada Y, Watanabe H (2007) A framework for measuring and evaluating program source code quality. In: Proceedings of the 8th International Conference on Product-Focused Software Process Improvement, PROFES. Springer, pp 284–299
8. Singh AP, Tomar P (2014) Estimation of component reusability through reusability metrics. Int J Comput Electr Autom Control Inf Eng 8(11):1965–1972
9. Sandhu PS, Singh H (2006) A reusability evaluation model for OO-based software components. Int J Comput Sci 1(4):259–264
10. Chidamber SR, Kemerer CF (1994) A metrics suite for object oriented design. IEEE Trans Softw Eng 20(6):476–493
11. Zhong S, Khoshgoftaar TM, Seliya N (2004) Unsupervised learning for expert-based software quality estimation. In: Proceedings of the Eighth IEEE International Conference on High Assurance Systems Engineering, HASE'04, pp 149–155
12. Kaur A, Monga H, Kaur M, Sandhu PS (2012) Identification and performance evaluation of reusable software components based neural network. Int J Res Eng Technol 1(2):100–104
13. Manhas S, Vashisht R, Sandhu PS, Neeru N (2010) Reusability evaluation model for procedure-based software systems. Int J Comput Electr Eng 2(6)
14. Kumar A (2012) Measuring software reusability using SVM based classifier approach. Int J Inf Technol Knowl Manag 5(1):205–209

15. Cai T, Lyu MR, Wong KF, Wong M (2001) ComPARE: a generic quality assessment environment for component-based software systems. In: Proceedings of the 2001 International Symposium on Information Systems and Engineering, ISE'2001
16. Bakota T, Hegedűs P, Körtvélyesi P, Ferenc R, Gyimóthy T (2011) A probabilistic software quality model. In: 27th IEEE International Conference on Software Maintenance (ICSM), pp 243–252
17. Papamichail M, Diamantopoulos T, Chrysovergis I, Samlidis P, Symeonidis A (2018) User-perceived reusability estimation based on analysis of software repositories. In: Proceedings of the 2018 IEEE International Workshop on Machine Learning Techniques for Software Quality Evaluation, MaLTeSQuE. Campobasso, Italy, pp 49–54
18. Samoladas I, Gousios G, Spinellis D, Stamelos I (2008) The SQO-OSS quality model: measurement based open source software evaluation. Open source development, communities and quality, pp 237–248
19. Hegedűs P, Bakota T, Ladányi G, Faragó C, Ferenc R (2013) A drill-down approach for measuring maintainability at source code element level. Electron Commun EASST 60
20. Ferreira KAM, Bigonha MAS, Bigonha RS, Mendes LFO, Almeida HC (2012) Identifying thresholds for object-oriented software metrics. J Syst Softw 85(2):244–257
21. Alves TL, Ypma C, Visser J (2010) Deriving metric thresholds from benchmark data. In: Proceedings of the IEEE International Conference on Software Maintenance, ICSM. IEEE, pp 1–10
22. Foucault M, Palyart M, Falleri JR, Blanc X (2014) Computing contextual metric thresholds. In: Proceedings of the 29th Annual ACM Symposium on Applied Computing. ACM, pp 1120–1125
23. Shatnawi R, Li W, Swain J, Newman T (2010) Finding software metrics threshold values using ROC curves. J Softw: Evol Process 22(1):1–16
24. Bay TG, Pauls K (2004) Reuse Frequency as Metric for Component Assessment. Technical report, ETH, Department of Computer Science, Zurich. Technical Reports D-INFK
25. SonarQube platform (2016). http://www.sonarqube.org/. Retrieved: June 2016
26. Papamichail M, Diamantopoulos T, Symeonidis A (2016) User-perceived source code quality estimation based on static analysis metrics. In: Proceedings of the 2016 IEEE International Conference on Software Quality, Reliability and Security, QRS. Vienna, Austria, pp 100–107
27. Borges H, Hora A, Valente MT (2016) Predicting the popularity of github repositories. In: Proceedings of the The 12th International Conference on Predictive Models and Data Analytics in Software Engineering, PROMISE 2016. ACM, New York, NY, USA, pp 9:1–9:10
28. ARiSA - Reusability related metrics (2008). http://www.arisa.se/compendium/node38.html. Retrieved: September 2017
29. SourceMeter static analysis tool (2017). https://www.sourcemeter.com/. Retrieved: November 2017
30. Doane DP, Seward LE (2011) Measuring skewness: a forgotten statistic. J Stat Educ 19(2):1–18
31. Dimaridou V, Kyprianidis AC, Papamichail M, Diamantopoulos TG, Symeonidis AL (2017) Towards modeling the user-perceived quality of source code using static analysis metrics. In: Proceedings of the 12th International Conference on Software Technologies - Volume 1, ICSOFT. INSTICC, SciTePress, Setubal, Portugal, pp 73–84
32. Dimaridou V, Kyprianidis AC, Papamichail M, Diamantopoulos T, Symeonidis A (2018) Assessing the user-perceived quality of source code components using static analysis metrics. In: Communications in Computer and Information Science. Springer, page in press

Part V
Conclusion and Future Work

Chapter 11
Conclusion

The main scope and purpose of this book has been the application of different methodologies in order to facilitate the phases of software development through reuse. In this context, we have applied various techniques in order to enable reuse in requirements and specifications, support the development process, and further focus on quality characteristics under the prism of reuse.

In Part II, we focused on the challenges of modeling and mining software requirements. Our methodology in requirements modeling provides a solid basis as it provides an extensible schema that allows storing and indexing software requirements. We have shown how it is possible to store requirements from multimodal formats, including functional requirements and UML diagrams. Our proposal of different methodologies for mining software requirements further indicates the potential of our ontology schema. The applied mining approaches have been shown to be effective for enabling reuse at requirements level. Another important contribution at this level is the mechanism that can directly translate requirements from different formats into a model that can subsequently serve as the basis of developing the application. This link, which we have explored in the scenario of RESTful services, is bidirectional and, most importantly, traceable.

In Part III, we focused on the challenges of source code mining in order to support reuse at different levels. We have initially shown how source code information can be retrieved, mined, stored, and indexed. After that, we confronted different challenges in the context of source code reuse. We first presented a system for software component reuse that encompasses important features; the system receives a query in an interface-like format (thus allowing easily to form queries), and retrieves and ranks software components using a syntax-aware model. The results are also further analyzed to assess whether they comply with the requirements of the developer (using tests) as well as to provide useful information, such as their control flow. Our second contribution has been in the area of API and snippet mining. The main methodology, in this case, addresses the challenge of finding useful snippets for common

© Springer Nature Switzerland AG 2020

T. Diamantopoulos and A. L. Symeonidis, *Mining Software Engineering Data for Software Reuse*, Advanced Information and Knowledge Processing, https://doi.org/10.1007/978-3-030-30106-4_11

developer queries. To do so, we have retrieved snippets from different online sources and have grouped them in order to allow the developer to distinguish among different implementations. An important concept presented as part of this work has been the ranking of the implementations and of the snippets according to developer preference, a useful feature that further facilitates finding the most commonly preferred snippets. Finally, we have focused on the challenge of further improving the selected implementations (or generally one's own source code) by finding similar solutions in online communities. Our contribution in this context has been a methodology for discovering solutions that answer similar problems to those presented to the developer. An important aspect of our work has been the design of a matching approach that considers all elements of question posts, including titles, tags, text, and, most importantly, source code, which we have shown to contain valuable information.

In Part IV, we focused on source code reusability, i.e., the extent to which software components are suitable for reuse. Initially, we constructed a system to illustrate the applications of reusability assessment. In specific, our system focuses on the challenge of component reuse to indicate how developers need to assess software components not only from a functional perspective but also from a quality perspective. As a result, we have proposed a methodology for assessing the reusability of software components. Our methodology is based on the intuition that components that are usually preferred by developers are more probable to have high quality. As part of our contribution to this aspect, we have initially shown how project popularity (preference by developers) correlates with reusability and how these attributes can be modeled using static analysis metrics. Our models have been shown to be capable of assessing software components without any expert help.

As a final note, this book has illustrated how the adoption of reuse practices can be facilitated in all phases of the software development life cycle. We have shown how software engineering data from multiple sources and in different formats can be mined in order to provide reusable solutions in the context of requirements elicitation and specification extraction, software development, and quality assessment (both for development and for maintenance). Our methodologies cover several different scenarios but, most importantly, offer an integrated solution. Our research contributions can be seen as the means to create better software by reusing all available information. And, as already mentioned in the introduction of this book, creating better software with minimal cost and effort has a direct impact on the economical and social aspects of everyday life.

Chapter 12
Future Work

As already mentioned in the previous chapter, our contributions lie in different phases of the software engineering life cycle. In this chapter, we discuss the future work with respect to our contributions as well as in regard to the relevant research areas.[1] We initially analyze each area on its own by connecting it to the corresponding part of this book, and then focus on the future work that can be identified for the field as a whole.

Future work on requirements modeling and mining (Part II) lies in multiple axes. Concerning the modeling part, the main challenge is to create a model that would encompass the requirements elicitation process as a whole. Such a model could include requirements, stakeholder preferences, non-functional characteristics of projects, etc. Our effort in this context can be considered a solid first step as it effectively stores and indexes some of this data, while it is also suited for employing mining techniques with the purpose of reuse. Similarly, and concerning also the requirements mining part, an important future direction is the support for different formats of requirements (e.g., user stories, different diagram types, etc.) and the application of mining techniques to facilitate requirements reuse as well as specification extraction. In the context of specification extraction, another important point to be explored in future work is that of connecting requirements to implementations. Our work in this aspect, which is focused on RESTful services, serves as a practical example and can provide the basis for further exploring this connection.

Concerning source code mining (Part III), the future research directions are related to the different challenges discussed. One such challenge is component reuse, for which we have proposed a methodology that assesses whether the retrieved

---

[1]An important note for the reader at this point is that highly specific future proposals for our techniques are provided at the end of each chapter. As a result, we focus here on the future work for the areas in a rather abstract level.

© Springer Nature Switzerland AG 2020

T. Diamantopoulos and A. L. Symeonidis, *Mining Software Engineering Data for Software Reuse*, Advanced Information and Knowledge Processing, https://doi.org/10.1007/978-3-030-30106-4_12

components are functionally suitable for the query of the developer. Interesting directions in this context include the support for larger multi-class components as well as the automated dependency extraction for each component. Ideally, we would want a system that could retrieve online source code, check if it is suitable for the corresponding requirements, and integrate it by taking into account the source code of the developer along with any dependencies to third-party code. Similarly, for the API/snippet reuse scenario, a future direction would be to integrate the snippet into the source code of the developer, possibly after employing summarization techniques to ensure that it is optimal. Finally, an interesting extension with respect to the challenge of finding existing solutions in online communities would be the incorporation of semantics. This way, one would be able to form a bidirectional semantic connection between a query that practically expresses some functional criteria and a relevant snippet. Such connections could be used for facilitating API/snippet discovery or even for automated documentation or extraction of usage examples from libraries.

The research directions for quality assessment (Part IV) are also multiple. At first, it would be interesting to further explore the incorporation of quality metrics into source code reuse systems. The main purpose of such an embedding would be to provide the developer with the best possible solution, one that would be functionally suitable for the given query and at the same time would exhibit high quality and/or possibly high preference by the developer community. In this context, it would be useful to investigate how different quality metrics and models can be used to assess source code components. Having the reusability assessment methodology provided in this book as a starting point, one could further explore the relation between developer preference and reusability in multiple ways. Developer preference measurements could be extended by measuring the actual reuse rate of each software library, a statistic that could be obtained by mining the imports of online repositories. Furthermore, the indicated relation could be explored in different domains; for instance, one would probably expect that general-purpose libraries are reused more often than machine learning libraries, which indicates that their preference has to be measured using a different basis.

Finally, we may say that in this book we have explored the concept of software reuse as a driving axis for software development. An interesting extension that spans over the whole process would be to model and index components of different granularity, along with their requirements, source code, and all other relevant information, and, through semantics, manage to provide an interface for finding reusable solutions in different levels (e.g., requirements, source code, documentation, etc.). Given traceable links between the data of the components, one would be able to develop software in a reuse-oriented manner, while the integrated solutions would effectively propagate when required in order to produce a clean, efficient, and maintainable product. Thus, as a final note, we may view this work as a solid first step toward this direction.

Printed in the United States
by Baker & Taylor Publisher Services